AF900852

Desert Weeds

Garry Rogers

Desert Weeds

Personal Narrative on Botanical First Responders

Springer

Garry Rogers
Agua Fria Open Space Alliance, Inc.
Humboldt, AZ, USA

ISBN 978-3-030-45853-9 ISBN 978-3-030-45854-6 (eBook)
https://doi.org/10.1007/978-3-030-45854-6

© Springer Nature Switzerland AG 2020
This work is subject to copyright. All rights are reserved by the Publisher, whether the whole or part of the material is concerned, specifically the rights of translation, reprinting, reuse of illustrations, recitation, broadcasting, reproduction on microfilms or in any other physical way, and transmission or information storage and retrieval, electronic adaptation, computer software, or by similar or dissimilar methodology now known or hereafter developed.
The use of general descriptive names, registered names, trademarks, service marks, etc. in this publication does not imply, even in the absence of a specific statement, that such names are exempt from the relevant protective laws and regulations and therefore free for general use.
The publisher, the authors, and the editors are safe to assume that the advice and information in this book are believed to be true and accurate at the date of publication. Neither the publisher nor the authors or the editors give a warranty, expressed or implied, with respect to the material contained herein or for any errors or omissions that may have been made. The publisher remains neutral with regard to jurisdictional claims in published maps and institutional affiliations.

This Springer imprint is published by the registered company Springer Nature Switzerland AG
The registered company address is: Gewerbestrasse 11, 6330 Cham, Switzerland

For Marc and Adam

Preface

While I studied at Arizona State University, my boys and I sometimes walked to campus to visit the library. We lived north of the Salt River, a once perennial desert stream dams and diversions had turned into a dry riverbed. The ASU campus lies south of the river and west of our apartment. The river banks and bed gave us a quieter walk with less automobile exhaust for most of the distance to the school. We joined the river a few blocks from our apartment and exited when we neared the campus. That two-mile hike was one of our favorite times together. The banks and channel held scattered and flood battered Cottonwood trees, Mesquite, Seepwillow, Acacia, clusters of Squawbush, brilliant Brittlebush, thickets of Sunflowers, Mullein spikes, feathery Storksbill, curly Heliotrope, other weeds, lots of bare ground, and a million stones smoothed and rounded by the river.

Between rock throwing attacks on old tires and washing machines, we saw birds, rabbits, and lizards, but no people. During our walk down the river to the urban campus of 40,000 students and thousands of staff, we never encountered another person. Not marketed for recreation and perhaps considered a wasteland unworthy of a visit, the Salt River's dry riverbed was a weed-dominated wildlife refuge running through the heart of a large city.

After graduation, my research on fire ecology and historical vegetation change brought me into frequent contact with weeds. I learned weeds grow faster, produce more seeds, and spread farther faster than other plants. In a single growing season, weeds colonize and cover sites laid bare by fires and floods. Some such as the thistles and milkweeds spread on the lightest breeze, others have thorny seedpods that hook a ride on anything brushing by, and some develop extensive root systems that spread and send up new shoots far from the parent plant.

The ability of weeds to reach and thrive in damaged and often-inimical environments suggests they may be the principal survivors of the detrimental effects of human overpopulation, habitat destruction, pollution, and climate change. As the most aggressive invaders spread, even the diversity of the weed vegetation declines. A quiet time of missing wildlife and vast weedlands could come, but this is what I think will happen to the weeds: Short-lived arthropods will soon evolve to prey on the introduced invasive weeds. Monotypic stands of the invaders will become patchy as the arths increase. Other weeds will find the openings and bring along their specialized consumers. And jumping ahead a bit, we can expect evolution will produce predators for the arthropods and will produce slow-growing, long-lived, pest-resistant, shade-casting plants. One day, desert grasslands, shrublands, woodlands, rodents, ungulates, and top predators making the most of local climate and soil conditions will return.

I wrote this book to clarify the nature of the weeds growing in natural desert and dryland landscapes. I wanted to answer these four questions:

1. How do weeds, particularly desert weeds, differ from other plants? For this answer, I considered the form, phenology, reproduction, and dynamic characteristics of the weeds.
2. Which weeds are present on the 20-acre patch of dry land where I live? Answering this question requires identifying the weeds. Illustrations and descriptions of weeds fill most of this book's pages. Most of the weeds growing here are present in deserts and drylands worldwide. The literature contains answers to the third question:
3. Are any desert weeds a valuable resource for wildlife and people? I derived the answer for this question from the published literature on:

   (a) soil and watershed responses to weeds,
   (b) wildlife uses of weeds for food and cover,
   (c) human uses of weeds for food and medicine.

   It is impossible to consider the potential value of weeds without acknowledging the destructive nature of some of them. A small proportion of the weeds people introduce to new regions become invasive and though they were probably beneficial in their homelands, in their new homes they replace native species and destroy ecosystems.
4. How will climate change and the impacts of the growing human population affect weeds? Will weeds become more abundant?

Humboldt, AZ, USA Garry Rogers

Acknowledgements

I am grateful for discussions with Ray and Jeanne Turner, Julio Betancourt, Jeff Steele, Denise Rogers, and members of the Yavapai Weed Management Area. Andrew Salywan, Ed Northam, Jolanta Sokol, and Patti Fenner contributed information and suggestions. Librarians Jeff Franklin, Gayle Braun, and staff of the Dewey-Humboldt Public Library helped find materials for the literature review.

Denise Rogers, Joanne Oellers, David Stith, Mary Franks, Tom Chester, and anonymous reviewers read and commented on the text. They were not proofreading, but they pointed out some of my errors.

I could not have assembled this book without the generosity of the illustrators and photographers who made their work available for download from the Internet, and those lovely souls who scanned and uploaded hundreds of old paintings by botanical artists. I am thankful for the University of Arizona Press permission to reproduce the excellent botanical drawings by Lucretia Breazeale Hamilton. Names of other artists and photographers are in the image captions and in a separate index.

Botanists and herbarium staff at Arizona State University, New York Botanical Garden, University of Arizona, and University of Utah gave generous help and encouragement. What I know of weeds and other plants includes a small part of their knowledge brushing off as I passed by.

Marc and Adam Rogers accompanied me on many field explorations and happily challenged my speculations. All field scientists should have such assistants.

In 1997, my wife, daughter, and I moved down from the eastern hills of Lonesome Valley to the Agua Fria River floodplain in the valley floor. The river, floodplain ponds, and riparian forest dominate this place, so we named it Coldwater Farm (CWF). Experience here has enriched my knowledge of weeds and has clarified a complex element of nature that was truly beneficial before we humans began carrying invasive species across the oceans, deserts, and forests.

Contents

Chapter 1
The Weed Problem

About 10% of the weeds introduced to new habitats can compete with and replace native plants. These invasive species cause a decline of diversity, productivity, and stability, they disrupt ecosystem processes, and they even alter the soil microbiome. During the past few centuries, people have introduced invasive weeds to every ecosystem on the planet. A few Eurasian weeds that reached the interior of the Western USA during the 1600s spread along trails and roads across the land. Mid-nineteenth century explorers reported finding invasive weeds in remote mountain canyons visited only by loggers (Watson 1874). Today the same weeds dominate many millions of acres. In 2007, the U.S. Bureau of Land Management (BLM) the largest landholder in the Western USA had this to say about the spreading weeds.

> There has also been a fourfold increase in invasive weed populations since 1985. Wildfires, drought, and invasive weeds are causing a steady degradation of soils, water quality and quantity, native plant communities, wildlife habitat, wilderness values, recreational opportunities, and livestock forage. The change has been so complete that only a sketchy picture of predisturbance conditions remains (BLM 2007, p. 30).

In many places, broad expanses of native grasslands, shrublands, and woodlands have become weedlands (e.g., Burcham 1970; Heady 1977; Mack 1981; Babbitt 1998; Simberloff 2013). In desert regions worldwide, human introduced weeds have usurped the land. They have changed the soils and replaced the original plants and animals. Their range has become so great the cost to replant the original vegetation is too great for society to bear.

© Springer Nature Switzerland AG 2020

G. Rogers, *Desert Weeds*, https://doi.org/10.1007/978-3-030-45854-6_1

Chapter 2
The Lives of Weeds

Weeds appear after a disturbance damages or removes plants and exposes the soil. Storms, fires, and seismic events create special environments for evolution of plants capable of growing on exposed soil in bright sunlight. Plants quick to disperse their seeds to disturbed sites and capable of withstanding the drying effect of full sunlight can grow free of interference from competitors. Competitors will come however, so rapid growth and reproduction are essential for bare-soil specialists. Except for the spreading roots of a few species and some short-lived trees, weeds die after 1 or 2 years. Even among weeds with long-lived roots, the parts we see, the shoots, are short-lived. Thus, natural disturbances create conditions for selection of an elite group of short-lived quick-maturing species with heavy seed production and great dispersal mechanisms.

Most weeds grow poorly in shade. Researchers have found weed success varies along a continuum from full sun to dense shade. Many can grow and reproduce in light shade such as cast by tall deciduous trees without a shrub understory. But only a few species such as shepherd's purse (*Capsella bursa-pastoris*) can reproduce in deeper shade (Yasin et al. 2019). No weeds grow in dense shade such as cast by multi-layered canopies or single-layered canopies of overlapping leaves such as those of the plums.

In deserts, drought-tolerant weeds flourish in disturbed upland slope habitats, and moisture-dependent species follow the banks of streams and springs (Fig. 2.1). However, weeds grow rapidly, and the moisture-dependent species often spread to drier habitats during brief benign periods. Summer cypress (*Kochia scoparia*), for example, can mature and produce a single spike of seeds at a height of three centimeters. When enough moisture is available, the plants reach up to four meters in height and produce many seed laden spikes. The adventurous moisture lovers may disappear when average conditions return, but many of them leave behind persistent seeds and roots ready to continue advancing in future wet periods.

© Springer Nature Switzerland AG 2020

G. Rogers, *Desert Weeds*, https://doi.org/10.1007/978-3-030-45854-6_2

Fig. 2.1 Rabbitfoot grass (*Polypogon monspeliensis*). This moisture-loving grass may have originated in Europe, but it became a global cosmopolitan so early in human history its origin is uncertain. Rabbitfoot grass can spread through dry upland sites by following intermittent stream channels and roadside ditches watered by runoff from impermeable road surfaces (photo © GR)

During the normal course of events, large, mat-forming, shade-casting, long-lived species of the desert grasslands, shrublands, and woodlands spread back into disturbed sites and replace the weeds. The long-lived plants add physical diversity, build complex nutrient cycles, and form interacting groups of long-lived species that maximize production from incoming solar radiation, local minerals, and local climate. These mature communities are evolution's most advanced creations.

In their rapid response to vegetation disturbance by construction, fires, floods, and foraging animals, weeds resemble human first responders. Weeds cover exposed soil and provide protection from erosion. Following geomorphic events such as landslides, weeds bind sand and gravel and provide the organic debris needed to rebuild the soil. Were we to remove all the weeds, injuries to ecosystems would heal slowly or not at all. Barren surfaces could appear and expand, flooding and erosion would increase, and productivity would fade.

However, when people introduce weeds to places beyond their normal range, problems can develop. Freed from home range consumers, predators, and competitors, a few introduced weeds will spread across local habitats, use nutrients needed by local plants, disrupt species interactions, break food chains and nutrient cycles, and destroy original ecosystems.

2.1 Weed Dispersal

Weeds are the travelers of the plant world. Unlike apples, weed fruits often fall far from the tree. The greater dispersal capability of weeds correlates with their short lives (Beckman et al. 2018). The principal dispersal vectors are birds, wind, water, and animals walking and driving. Weed seedlings appear when fire or flood destroys vegetation and they appear beside every road, around every tree and shrub, and beside every fence and walkway railing where birds perch.

Weeds also disperse using underground rhizomes, surface stolons, and vines. These plant organs can creep along with surprising speed, but none of these attached parts attains the velocity achieved by the seeds of the weeds whose seedpods explode when they become dry. These botanical bombardiers propel their seeds surprising distances. The bursting seedpods of Creeping Woodsorrel (*Oxalis corniculata*), a tiny plant common on moist peripheries of fields and streams, shoot seeds up to 10 ft. (Fig. 2.2).

Dispersal over large distances is rare; most propagules—the seeds and live stems and roots—remain near their parents even when they have physical features that facilitate dispersal (Ridley 1930; van der Pijl 1972; Sheldon and Burrows 1973). One study of four weed species found most seeds fell within 20in of the parent plant and 100% of the seeds of three species were within 40in of the parent (Marchand and Roach 1980). Though such short distances are common, dispersal vectors will carry seeds over and around roads, fields, rivers, and mountains.

Fig. 2.2 Yellow flowers of Creeping Woodsorrel with grass and dandelions (photo © GR)

Many ornamental or food-producing plants introduced to new locales by people have produced unintended consequences. Some of these plants were weeds. Just as introducing a mammal without its predators allows destructive population growth, introducing a weed without its consumers allows runaway invasion.

2.2 Human Introduced Weeds

Birds and wind can carry weed propagules long distances, but it was not until the 1500s when sailing ships carried passengers and cargo between the Old World and the New destructive weed invasions could happen. Thousands of species of weeds began and continue crossing the oceans mixed with grain and stowed away on board ships and planes (Janzen 1967; Crosby 1952). After 500 years of constant human contribution to dispersal, the reservoirs of weedy colonists are not empty. New weeds are still crossing the oceans and entering ecosystems far from home.

Ships crossing between the Old and New world often dump ballast on land and at sea that contains weed propagules. Weeds also arrive mixed with grain and seeds of horticultural plants, stuck in the fur of livestock, the fabric of clothing, and in the treads of tires (Shao et al. 2019). Robert Frenkel once analyzed the debris traps of the drains at a car wash and found the seeds and vital rhizomes of over 400 plant species (Frenkel 1970).

Most transcontinental weed dispersal has been from the Old World to the New. Weed scientists believe Eurasian weeds evolved more efficient traits than New World weeds because there is greater variety of environments and selective pressures across the larger Eurasian continent with its broad connection to Africa (Di Castri 1989). The longer Old World history of human occupation and disturbance created opportunities for evolution of more advanced weed abilities. For example, no North American weeds can match the overwhelming success of cheatgrass (*Bromus tectorum*), redstem filaree (*Erodium cicutarium*), red brome (*Bromus rubens*), and others that in various combinations have spread across the arid Western United States (Young et al. 1971; Mack 1989; Reid et al. 2008; Turner et al. 2010).

Though less frequent, New World weeds have dispersed to the Old. For example, Shao et al. (2019) report Lance-leaf Sage (*Salvia reflexa*) has several small colonies near a grain transfer site in Northeast China. And there are records of the weed in Australia, Europe, New Zealand, and South Africa.

Horticulturalists and range scientists have imported plants that have damaged ecosystems far beyond the places of their introduction. For example, nurseries everywhere sell yellow-flowered Toadflax (*Linaria spp.*), beautiful plants of Eurasian origin (Fig. 2.3). The plant's roots delve deep and may send up shoots outside flowerbeds, in neighbor's yards, and across the road. Controlling Toadflax with regular pulling and mowing is easy in gardens and yards, but not so easy when it spreads across native shrublands and woodlands. The plants have established their deep root systems across thousands of acres of arid vegetation in the region surrounding CWF. Today, Dalmatian Toadflax has spread so far eradication would take an enormous investment. The later section on weed control discusses what it takes to get such an investment. Meanwhile, like an untreatable disease, this pretty flower continues spreading and robbing native plants of moisture and nutrients.

Fig. 2.3 Dalmatian Toadflax (*Linaria dalmatica*) (photo © GR)

Lehmann Lovegrass (*Eragrostis lehmanniana*) provides an example of a plant land managers imported for its utility but were surprised when the plant spread from the roadsides into native desert grasslands. Around 1930, Arizona land managers heard about an Old World plant that would feed livestock and would colonize and grow on land cows had stripped of other palatable plants. The managers imported seeds and planted them on overgrazed lands. Regional transportation engineers planted the grass on eroded roadsides. The grass grew well, and the cows were happy and the roadsides stopped eroding. Soon, however, ecologists realized the plant was invasive (Cox 1992). By 2010 when all planting in the region around CWF stopped, the plant had replaced almost half a million acres of desert grassland. It had crowded out native grasses and formed pure stands with reduced plant and animal diversity.

Another imported invasive weed is Sweet Resinbush (*Euryops subcarnosus*). This little shrub has bright yellow flowers and a distinct odor some of us find offensive. Grazing managers with the U.S. Department of Agriculture planted this South African perennial in an experimental plot in Arizona in 1935 with the mistaken belief it was a desirable forage plant for cattle (McReynolds 2008). Soon realizing cattle would not eat the stinky little bush, the investigators lost interest. Untended and forgotten, the plant persisted in its original plots until almost 30 years later it spread. Before weed scientists realized what was happening, the plant had spread across thousands of acres. At present, Sweet Resinbush colonists are pushing north along roads, transmission corridors, and regional desert rivers. It will reach CWF soon. I think it will be easier to control than our worst smelly plant, Blue Mustard (*Chorispora tenella*), but prevention, discussed below, would preserve olfactory contentment here at CWF.

There are examples of introductions with unexpected consequences to fill several books. The image below (Fig. 2.4) shows an example of an introduced species land managers planted (Crested Wheatgrass, *Agropyron cristatum*) is being replaced by another introduced species (Cheatgrass, *Bromus tectorum*).

Accidental and intentional introductions of weeds like Lehman Lovegrass, Sweet Resinbush, and Crested Wheatgrass continue today. Ranchers still dream of finding a plant that protects the soil while producing a never-ending supply of abundant nutritious forage. They might sooner find they can use cold fusion to grow money on trees.

Fig. 2.4 Cheatgrass vs. Crested Wheatgrass. I drove by this place many times admiring the dense Pinyon-Juniper (*Pinus edulis, Juniperus osteosperma*) woods with its mid-level layer of interior chaparral shrubs, lower layer of Sagebrush (*Artemisia tridentata*), and its lowest layer of herbs and biological soil crust. The Eurasian invader Cheatgrass (*Bromus tectorum*) wedged its way into the woods and helped fuel a fire that killed the trees. Land managers planted Crested Wheatgrass across the area (green clumps in the foreground), but it appears to be losing to Cheatgrass (reddish grass) (photo © GR)

2.3 Disturbance and Weed Establishment

In most instances, weed germination occurs on disturbed sites, those places where weather, human activity, or seismic events have thinned or removed the vegetation. Some invasive weeds appear to invade undisturbed natural vegetation but determining this is difficult. Human disturbances have affected every square inch of the world's deserts. There are sites where natural barriers or fences exclude livestock grazing, but the weedy land surrounding these sites can shower them with such a heavy rain of weed propagules weeds colonize tiny transient openings. These protected sites also serve as magnets for wildlife and suffer heavy use by small animals that pass through barriers.

Before people arrived, natural disturbances were smaller and less frequent, and weeds were a small part of regional floras. Even after early immigrants reached the New World 40,000 years ago, disturbances were not as frequent or widespread as they became when Europeans arrived. The growth and spread of human civilization added many new disturbances and sites available for weed colonization. The principal disturbances are construction, farming, domestic livestock grazing, roads, and human-caused fires. Droughts, storms, and floods arising from climate change are adding even more demands for nature's EMTs.

Some introduced weeds can spread into areas minimally disturbed. When we transport plants to new regions, we release them from the natural limitations of their homeland. We also expose them to new competitors, predators, and environments. The new conditions limit the growth and prevent the spread of most foreigners, but not all. Some of the weeds introduced to North America from Eurasia and Africa have spread across the land. Their more advanced competitive abilities developed in the larger, longer-human-disturbed lands let them absorb moisture and nutrients faster than native species, let them repel natives with toxic secretions, and let them fuel fires their seeds can survive. These latter species, which I call the fuelweeds, are the greatest of all threats to desert ecosystems.

In this time of extreme human impacts on nature, a time some are calling the Anthropocene era of Earth history, pre-disturbance vegetation rarely has time to replace the weeds before another disturbance occurs. In every desert region, we see places where taller, longer-lived plants are missing, where weeds have become the permanent vegetation (Turner et al. 2010).

Chapter 3
Controlling Weeds

One of my childhood chores was weeding. By the age at which I entered school, I had become adept at the wide swings of the weed hook. I always felt there must be better ways to earn room and board than spending summertime battling the weeds. Then chemical companies invented herbicides and gave every child and land manager a magic wand. Wave a sprayer across the weeds and they shrivel and die. Now that was progress. Chemical manufacturers proclaimed the herbicides were safe and some of their representatives even claimed you could drink the pure undiluted chemical without injury (though I doubt anyone did it). Every year, government land managers, farmers, and homeowners spray millions of gallons across the land around the world.

Fig. 3.1 Native Blister Beetles (probably *Epicauta pardalis*) on native Wright Saltbush (*Atriplex wrightii*). This beetle has limited the advance of Palmer Pigweed (*Amaranthus palmerii*) at Coldwater Farm after its rapid spread during 2018 (photo © GR)

© Springer Nature Switzerland AG 2020

G. Rogers, *Desert Weeds*, https://doi.org/10.1007/978-3-030-45854-6_3

Natural weed control occurs as native long-lived plants return to disturbed sites and replace the weeds. It also occurs when native insects and other animals consume the weeds and limit their growth (Fig. 3.1). Sometimes, it has been possible to control a weed by importing insects and other consumers from its homeland.

3.1 Early Detection, Monitoring, and Rapid Response

Early detection of invasive weeds is critical. While most introduced species never spread into native vegetation, some will go through a quiet phase lasting years or even decades and then began to spread. Once the invasion begins, large colonies with many distant satellite populations can form in a few years. Because of the uncertain survival of new weeds, the first response to most species is establishing a monitoring schedule. However, immediate response is necessary for weeds that begin spreading or that are known invaders in similar regions. Detailed information for weed response planning is available online from the North American Invasive Species Management Association. Another useful site, the Center for Agriculture and Biosciences International (CABI), provides news and invasive species information indexed by country, and a few subnations. The site addresses are in the Internet Sources list following the Sources.

3.2 Strategic Concepts

3.2.1 Adaptive Management

Control treatments applied to new weeds don't always work. Many interacting biological and physical conditions make weed response to treatment so unpredictable weed managers must view treatments of new weeds as experiments and be prepared to try different tactics.

3.2.2 Justification for Invasive Weed Management

Public participation in weed management helps with early detection and it is often necessary for getting financial support. However, convincing people to support weed control can be difficult. The field of Weed Science developed in response to the value of lost crop yields resulting from weed invasions. Comparable measures of the cost of grazed and wild-land invasions are more recent but are now well developed (Naylor 2000).

3.2.3 Prevention, Regulation, and Ecosystem Maintenance

Preventing weed invasions requires government policies and regulations that block weed imports by people, livestock, and equipment. Most places have land use policies for weed prevention that regulate livestock grazing, wood gathering, recreation, construction, and more. Often, however, weed managers cannot enforce the policies. The problem is the cost. Weed dispersal occurs in so many ways total prevention is an enormous expense. The best approach is to enlist public participation. For example, the Play, Clean, Go program cited below in the List of Relevant Websites publicizes the invasive species problem and enlists citizen participation in curtailing weed spread.

Another important part of weed prevention is ecosystem maintenance. Undisturbed native vegetation resists invasive species. Well-developed biological soil crusts (Fig. 3.4) for instance, are effective in blocking weed establishment (Slate et al. 2018). Maintenance focuses on preserving, and sometimes restoring, the diversity and stability of existing vegetation.

3.2.4 Satellite Populations

Weed populations grow by expanding their boundary and by long-distance dispersal to establish satellite populations. Since a single plant can produce thousands of seeds, tiny satellite populations of one or a few plants can serve as significant sources for further invasion. As the area of an infestation grows, the proportion of each plant's propagules falling among the plants within the occupied area will increase and the proportion falling outside the area will decline. Thus, the rate of expansion will decrease as the infested area grows. Because the rate of expansion is greater for small satellite populations, managers give treatment of satellite populations the highest priority.

3.2.5 Coordination with Neighbors

Coordinating prevention, inventory, and control with neighboring agencies and private landowners is a necessary part of weed management planning. Property boundaries do little to influence weed dispersal and establishment. Fences control livestock movements and stop wind-dispersed tumbleweeds, but they do not block birds and mammals transporting weeds in their gut or stuck to their bodies. Roads and trails provide access pathways for weed propagules carried by vehicles and people.

3.2.6 Control Vs. Eradication

Immediate eradication of invasive species is sometimes the most cost effective option. Annual costs to control weeds increases as new species arrive and existing species spread and establish satellite populations. Even though eradication of a widespread species is difficult and costly, it might be essential to preserve high-priority habitats.

3.3 Inventory

Inventories to discover and record the locations of invasive species rely on two approaches. The first is for continuous weed cover or partial cover of weeds scattered throughout native vegetation. It yields vegetation maps made using standard methods (Mueller-Dombois and Ellenberg 1974).

Restoring large sites with continuous weed cover is so difficult weed managers assign such sites to the lowest priority for treatment (see the next section). Inventories include these sites because they are often the first that invasive species entering a region will invade and because they serve as reservoirs of invasive species propagules.

The second method is for areas with scattered weed patches. It emphasizes the search for weed infestations. It creates narrative descriptions and sometimes includes small maps of weed occurrence. Weed patches occur around trees, along streams, roads, and fences, and in places where livestock gather.

3.4 Prioritizing Weed Species and Sites in Deserts

3.4.1 Prioritizing Species

Prioritization equips weed managers for immediate response when known invaders appear. The practice begins with searches for information on the ecology, distribution, and control of individual species. The information produced by the searches provides a first approximation of the relevant characteristics and control treatments for species in, or likely to reach a management area. Critical considerations for prioritizing species include potential for arrival, establishment, spread, and impact. Holm et al. (1977) provide detailed descriptions of the 76 weed species responsible for most of the world's agricultural losses. Larger weed-species data sets are available from the Global Register of Introduced and Invasive Species and the Invasive Species Compendium hosted by the Center for Agriculture and Biosciences International. Internet links for these and other Internet information sources follow the references.

The ideal prioritization systems integrate ecological knowledge of individual species with site conditions, and managers adjust it with monitoring information and observed results of treatments.

3.4.2 Prioritizing Sites

Hobbs and Humphries (1995) proposed ranking sites for weed control based on two factors, ecosystem value and level of site disturbance. Protected parks, wilderness areas, and critical habitats such as desert streams have more ecosystem value and should receive the greatest weed-control effort (Fig. 3.2). Weed control costs may exceed the financial limits of the local government. This is true of sites where disturbances have removed native vegetation and sites on which invasive weeds are dominant may not be within the financial limits of the local agency.

Fig. 3.2 Damaged streambank. Water diversions, livestock, human travel and settlement, and floods impact desert riparian habitats. The site in this photograph has high value despite the impacts because it is a perennial stream with a small willow-cottonwood riparian forest occupying the floodplain (photo © GR)

The next figure (Fig. 3.3) illustrates weed management responses for sites and species of varying priority. Its application follows thorough botanical inventories and resource assessments for weed management units defined by physical geography, land use, vegetation, and political boundaries. These are the basic spatial units for weed control.

Prioritization for Species and Sites with Management Actions

Weeds⇨ / Sites ⇩	Highest	High	Medium	Low
Highest (<1)	(0) Withdraw, eradicate, and restore.	(0) Withdraw, eradicate, and restore.	(2) Monitor, treat if expanding.	(3) Monitor, treat if expanding.
High (1)	(0) Withdraw, evaluate for eradication, & restoration.	(1) Withdraw, evaluate for eradication & restoration.	(2) Monitor, treat if expanding.	(3) Monitor, treat if expanding.
Medium (2)	(0) Withdraw, begin control treatment.	(2) Monitor, treat if expanding.	(3) Monitor.	(5) Monitor.
Low (3)	(1) Withdraw, begin control treatment.	(3) Monitor, treat if expanding.	(5) Monitor.	No Action.

Fig. 3.3 Priority sites and species. Numbers in the left column represent the minimum number of years between weed inventories. Highest-priority sites receive two or more visits each year because some weeds germinate throughout the growing season and others germinate during spring or during summer. Numbers in cells (not including the left column) show the recommended frequency in years for monitoring existing infestations. Zero (0) in the cells for highest and high priority sites and weeds show rapid response instead of monitoring. Seasonal timing of observations follows maturation of the weeds. The term "withdraw" refers to intervention to limit human disturbances. The cells at lower right indicate sites to monitor to detect appearances by high-priority species. Depending on resources and proximity to high-value sites, managers might eradicate such species (image © GR)

3.5 Treatment

Treatment tactics include chemical herbicides, flame and steam applicators, fabric or mulch ground cover, pulling, cutting, releases of predatory insects and microorganisms, planting native species, landscape burning, and combinations of these. Published information on individual weed responses to treatments is often available, but for many weeds, results of treatments vary with location, timing, and species.

The best way forward is careful documentation of treatments so failures and successes can inform future efforts.

Chemical weed control has become the major treatment tactic for irrigated farms, and glyphosate, the active ingredient in RoundUp has become the principal chemical. The amount of glyphosate used in the U. S. rose from 7,683,000 pounds in 1990 to 249,906,307 pounds in 2014. Use exploded because of the development of glyphosate resistant crops (Benbrook 2016).

Herbicides are not 100% effective. Regular application creates a new environmental force that selects for resistant members of treated populations. In some regions, most weeds have become resistant. Farmers have responded with heavier doses of herbicides, and with combinations of herbicides.

Claims the herbicides are harmful to wild animals and humans have been increasing. Reports of wildlife declines and human illness have led to some governments bans (Carson 1962; Jameson 2012). In the U. S., lawsuits are multiplying and bans are increasing (Gillam 2017). Consumers have begun avoiding fruits and vegetables treated with herbicides and this has fueled growth of chemical-free organic farming. However, industrial-style farming of corn, soybeans, and other large-scale crop plantings require chemical herbicides (Zimdahl 2010).

Non-chemical control alternatives are less effective than glyphosate because they do not provide complete cover, they require too much labor, or they have their own negative impacts. Though limited in comparison with glyphosate, the use of non-chemical methods is increasing (Jabran and Chauhan 2018; Brodie 2018). Thermal weed control, for instance, has less impact on nontarget plants and arthropods, and is effective for pre-emergent treatment of many weeds. However, it is harmful to soil microorganisms, and it cannot provide complete coverage once crops germinate.

3.6 Restoration

Healthy native vegetation resists weed invasions. However, restoring native vegetation by eradicating invasive weeds and preventing their return while aiding establishment of native species is difficult and costly. Weeds often leave behind persistent propagules, changed soil chemistry, and depleted soil microorganism communities. Reestablishing biological soil crusts (Fig. 3.4) is difficult because the process is so slow; it may require several decades (Cole 1990). Discovering and correcting the changes is so expensive we can restore only the highest priority sites.

Fig. 3.4 Intact biological soil crusts inhibit weed invasion. Crusts are common in North American deserts, but they are uncommon in Old World deserts (photo © GR)

Chapter 4
Weeds, Nature Conservation, and Global Warming

Naturalists have long considered invasive plants and animals to be the greatest dangers to nature after destruction of soil and vegetation by human digging, plowing, and building. Biologists and conservation organizations report that because of lost habitat, more than half of all wild animals are already gone (World Wildlife Fund 2016).

Over the past few decades, the effects of global warming have dominated concerns for nature and even for human civilization. According to the U.S. government, humans are causing substantial and growing climate changes (Wuebbles et al. 2017). For years, I did not believe global warming could destroy the land like construction of a road or building does. I was wrong; human-caused climate change is not limited to separate places on the land, warming will occur everywhere. Many weeds have evolved metabolic systems that function more efficiently at higher temperatures than most other plants. Known as C_4 and CAM, these adaptations will give weeds an advantage as Earth's climates warm and heat waves become more frequent. Earth's grasslands, shrublands, and forests will become weedlands with frequent fires, floods, dust storms, and eerie silence or monotonous calls of a few species of birds and insects.

During Earth's glacial episodes, ice-free enclaves called refugia sheltered groups of plant and animal species from physical destruction by spreading ice. Refugia acted as biodiversity reservoirs that resupplied life as the glaciers retreated. Global-warming refugia will include shady mountain canyons and riparian areas. However, extreme freezes, floods, droughts, fires, and heat waves will affect those areas, making invasive plants the most likely survivors and future dominants of the mountains and the plains.

© Springer Nature Switzerland AG 2020

G. Rogers, *Desert Weeds*, https://doi.org/10.1007/978-3-030-45854-6_4

Chapter 5
Coldwater Farm Habitats

The desert weeds illustrated in this book occupy a small farm in the Southwestern USA. At average altitude of 4480 ft., CWF covers 20 acres on the floor of a fault-block depression known as Lonesome Valley. The valley tilts south between the Bradshaw Mountains to the west and a line of low hills on the east. The farm includes upland, riparian, and lotic habitats near and within a small perennial stream known as the Agua Fria River (Fig. 5.1).

Fig. 5.1 Agua Fria River. Willow (*Salix*) and Cottonwood (*Populus fremontii*) dominate the riparian phreatophytic vegetation. Many other species including many weed species line the river. At CWF, a small closed-canopy forest fills the river floodplain (photo © GR)

© Springer Nature Switzerland AG 2020

G. Rogers, *Desert Weeds*, https://doi.org/10.1007/978-3-030-45854-6_5

Lonesome Valley is arid with hot summers and cold winters. Drought-tolerant trees, shrubs, and grasses dominate the valley floor and upland slopes. Phreatophytic trees and shrubs gather along the river and around ponds and springs. The valley lies between three of the North American deserts: The hot Sonoran Desert south, cool Mojave Desert west, and cold Great Basin Desert north. The weed flora of CWF combines elements of the three deserts and adjacent drylands but does not include species adapted to the hottest and coldest portions of the Sonoran and Great Basin Deserts.

When the Agua Fria River floods, the riparian vegetation bordering the stream slows water roaring down the main channel and spreads it across the flood plain. Despite the slowing, floods always rip up some plants and expose roots of others. Weeds colonize these injuries and hold the soil in place.

People removed the natural vegetation of Lonesome Valley long ago and planted irrigated crops and exotic pasture grasses. Water pumped from a well near the center of CWF maintained the farm's pastures. Only three cows lived on the farm from 1997 until the last one died in 2009. I fallowed the CWF pastures that year and weeds covered the exposed soil as the pasture grasses dried and died.

The stretch of the Agua Fria River running through CWF is a critical regional habitat. The open water supports amphibians, arthropods, birds, fish, lizards, mammals, snakes, and turtles. Hundreds of wildlife species spend all or part of their lives near the river. Yellow-billed Cuckoos and Southwest Willow Flycatchers, both on the U.S. Endangered Species List, nest beside the river (Rogers 2019). All the creatures living in the river or coming to it for water are components of Lonesome Valley's food webs stretching from the river to the mountains and from the soil to air above the trees.

Chapter 6
Drawings, Paintings, and Photographs

Plant identification is easier with drawings made by an experienced botanical illustrator than with photographs. An illustrator can emphasize the essential features distinguishing species. However, photographs are an inexpensive means to show plant colors and scenes with mixed species.

For each weed, I present the best graphic illustrations available. Many of the drawings are by Lucretia Breazeale Hamilton from the book *An Illustrated Guide to Arizona Weeds* by Parker (1972). I include them with the generous permission of the University of Arizona Press. Some of the photographs are by me and others have Creative-Commons licenses that allow reproduction with attribution to the photographer (CC BY 2-4 and BY-SA 2-4). I did not alter the photographs except as needed to fit them on the page and make them suitable for printing. Some of the drawings and photographs are from U. S. government web sites and are in the public domain. Paintings were available for some of the weeds. The ones I used are over 100 years old and are in the public domain. For all illustrations, I gave creator names in the captions and in an Index of Illustrators, Painters, and Photographers before the General Index.

You can find more works by the creators by entering their names or the names of the plants they depicted in the search box at Wikimedia Commons, Wikipedia, or Flickr Creative Commons. "GR" in a caption identifies photos by me. You can use my photos as long as you attribute them as "© Garry Rogers." Look up Creative Commons BY-SA 4.0 to read the license requirements.

© Springer Nature Switzerland AG 2020

G. Rogers, *Desert Weeds*, https://doi.org/10.1007/978-3-030-45854-6_6

Chapter 7
Plant Names

Plant names often present a challenge. Common names for most species vary from place to place, and scientific names intended to simplify communication can themselves change. Botanists often change plant names to add variants, combine species, or even move a species to a different genus or family (Turland et al. 2018). However, common names are maddening. Even related publications such as the medicinal plant books by Charles Kane and his inspiration, Michael Moore, use different common names for the same species.

The scientific naming system assigns related genera to families, related families to orders, related orders to classes, and related classes to kingdoms (plants, animals, etc.). This book includes the species, genus, and family names. To keep similar weeds together, I grouped the species description pages by family and I alphabetized the families by their scientific names.

The common names of many plants are older than the scientific names. If you want to learn more about the origin and history of common names, you can get a quick start with the book by Gledhill (1985). Most of the common names I use in this book are from *Arizona Weeds* by Parker (1972). The Weed Science Society of America (http://wssa.net) has a comprehensive list weed scientists consider the standard for botanical common names. However, getting everyone to use the same common names for plants is as unlikely as getting Americans to replace yards with meters.

© Springer Nature Switzerland AG 2020

G. Rogers, *Desert Weeds*, https://doi.org/10.1007/978-3-030-45854-6_7

Chapter 8
Desert Weed Identification

Drawings and photographs suffice to identify most species. For a few, however, measurements are necessary. You will need a 10x hand lens, a ruler, and a yardstick or retractable tape measure marked with standard (feet and inches) and metric (meters, centimeters, and millimeters) units.

8.1 Weed Species in this Book

This book includes drawings, photographs, paintings, and descriptions for 130 species I've found growing in CWF and 19 species botanists have found near the farm in habitats present on the farm.

Many of the weed species found at CWF are also present in Africa, Australia, Eurasia, South America, and many islands. Most of the weeds originated in Eurasia, but some are from the other continents, and some are natives of the region surrounding CWF. A few of the natives have spread to other continents, and others could do so. Mesquite, a native shrubby tree, for example, is a serious problem in Australia.

8.2 Arrangement

Each weed species has two pages. First pages have an illustration or photograph and a description with measurements for leaves, flowers, and seeds. Second pages include symbols indicating origin, prominent features, my perception of the weed, general remarks on the characteristics, distribution, and uses of the weeds, and additional pictures. The symbols appear in brackets following the plant name. For example, weeds whose dispersal mechanisms include sharp hooks or barbs receive a **P** (painful) and weeds difficult to eradicate receive a **T** (troublesome).

© Springer Nature Switzerland AG 2020

G. Rogers, *Desert Weeds*, https://doi.org/10.1007/978-3-030-45854-6_8

A: Annuals that sprout from seed.

B: Biennials sprout the first year from seed and the second year from roots.

P: Perennials sprout from roots or surviving aboveground structures.

A vertical line | separates the letters above from the ones below.

D: Delightful weeds to ignore or encourage because they are attractive, fragrant, or desirable for other reasons. For example, you may want to plant Milkweeds because they are a preferred host of Monarch Butterfly caterpillars.

E = Edible plants according to various sources, but especially the books by Moore (1989, 2003), Kane (2017), Tull (2013), and the website by Plants for a Future (2019).

F = Future. I have not seen these weeds in CWF, but they are present in in similar environments surrounding the farm.

I = Invasive. Some alien weeds can replace native vegetation and reduce the overall amount of wildlife food and cover. The worst of these get the letter **I**.

M = Medical plants according to the books by Moore (1989, 2003), Kane (2017), and others cited in the text.

N = Native weeds in North American deserts before people crossed the oceans and introduced weeds from other continents.

P = Obnoxious weeds with painful or irritating seeds or thorns including those that stick in animal fur (and Blue Mustard, because of its foul odor).

T = Troublesome. These weeds have persistent seeds or roots that make them difficult to eradicate.

Noxious weeds: Many countries, states, and provinces have lists of restricted weeds. Arizona, where CWF is, has a list with three categories: weed species to control (blast out of existence), species to quarantine whenever found, and species to block from entering the state. The third list has 54 species. However, it's like my Amazon wish list: Not possible under foreseeable budget limits. Most weed managers would agree all three lists are wish lists. Nevertheless, I marked the species on the first two lists as Z−1 and Z−2 occurring at CWF. What I should do about them is unclear. Their presence throughout the state makes it unlikely we will control or quarantine them as called for by the state's regulations.

Z−1 = Regulated pest in Arizona. This means the plant species, including viable plant parts (stolons, rhizomes, cuttings and seed, except seeds of agricultural and ornamental species), found within the state should be destroyed to prevent further infestation or contamination. Three CWF weeds are on this list.

Z−2 = Restricted pest in Arizona. This means the species, including viable plant parts (stolons, rhizomes, cuttings and seed, except agricultural and ornamental seed), found within the state shall be quarantined to prevent further infestation or contamination. Ten CWF weeds are on this list.

Chapter 9
Field Guide

© Springer Nature Switzerland AG 2020

G. Rogers, *Desert Weeds*, https://doi.org/10.1007/978-3-030-45854-6_9

9.1 Amaranth Family: AMARANTHACEAE

Fig. 9.1 Wright Saltbush (*Atriplex wrightii*)—AMARANTHACEAE. Annual reproducing from seed. Erect stems 2–4 ft. high. **Leaves**: *Silvery powder below, bright green above, edges wavy or toothed*, alternate, 2–8 cm long, 6–25 mm wide. **Flowers**: Male flowers in 2–12 in. leafless spikes on stem tops; female flowers in small clusters in leaf axils. **Seeds**: Pale brown, 1 mm wide, enclosed by pair of green or yellow fan-shaped bracts about 2 mm wide with 3–7 short teeth across the top. **a**. Male flowers. **b**. Cluster of three fruits (Fig. 9.1 from *An Illustrated Guide to Desert Weeds*, by Kittie F. Parker. © 1972 The Arizona Board of Regents. Reprinted by permission of the University of Arizona Press)

Wright Saltbush (*Atriplex wrightii*) [A|N]

The distribution maps from websites offering information on this and other weeds sometimes disagree. The Encyclopedia of Life (EOL 2019) and Flora of North America (eFloras 2019) report Wright Saltbush (Fig. 9.1) grows below 4000 ft. However, the SEINet (2019) map showing the species around CWF is correct; the plant does fine in CWF (altitude 4600 ft). The conflict might indicate pressed plants were not present in the herbaria the describers were using when they wrote the descriptions.

Information reported in many published plant descriptions comes from studies of pressed and preserved plant specimens stored in herbaria. If no one has collected a specimen from a particular area, a botanist describing a species' distribution might not report it occurs in that area. Though not perfect, relying on herbarium specimens to describe plants is an important practice. The stored specimens provide proof to support published descriptions.

Flora of North America (eFloras 2019) relies on Watson (1874) for the plant description. Male flowers at top of plant are shown in the figure below (Fig. 9.2).

Fig. 9.2 Wright Saltbush male flowers (photo © GR)

Fig. 9.3 Common Lambsquarters (*Chenopodium album*)—AMARANTHACEAE. One to seven feet high with mealy flowers and leaf bottoms. **Leaves**: Alternate, 1–5 in. long, and 12–50 cm wide. Higher leaves lanceolate and entire; lower leaves are egg or wedgeshaped with lobes at base and often with toothed margins. **Flowers**: Inconspicuous greenish flowers cluster on branch tips. **a**. Enlarged leaf and seed (**b**.) (Fig. 9.3 from *An Illustrated Guide to Desert Weeds*, by Kittie F. Parker. © 1972 The Arizona Board of Regents. Reprinted by permission of the University of Arizona Press)

Common Lambsquarters (*chenopodium album*) [A|E]

People have been cultivating and eating Lambsquarters (Fig. 9.3) for thousands of years. Asian and African farmers still plant it, but westerners consider it a useless weed. Cook the leaves or eat them raw, but limit your intake as they contain high levels of oxalic acid. The seeds are high in protein, vitamin A, calcium phosphorus, and potassium (U.S. Department of Agriculture Natural Resources Conservation Service 2019).

Lambsquarters readily invades crops and disturbed habitats. And if it makes seeds, it is difficult to eradicate—its seeds remain viable for up to 40 years.

This tough nutritious plant could be one of our major crops on the other end of the Anthropoce*ne* (Figs. 9.4 and 9.5).

Fig. 9.4 Lambsquarters flowering branch (© GR)

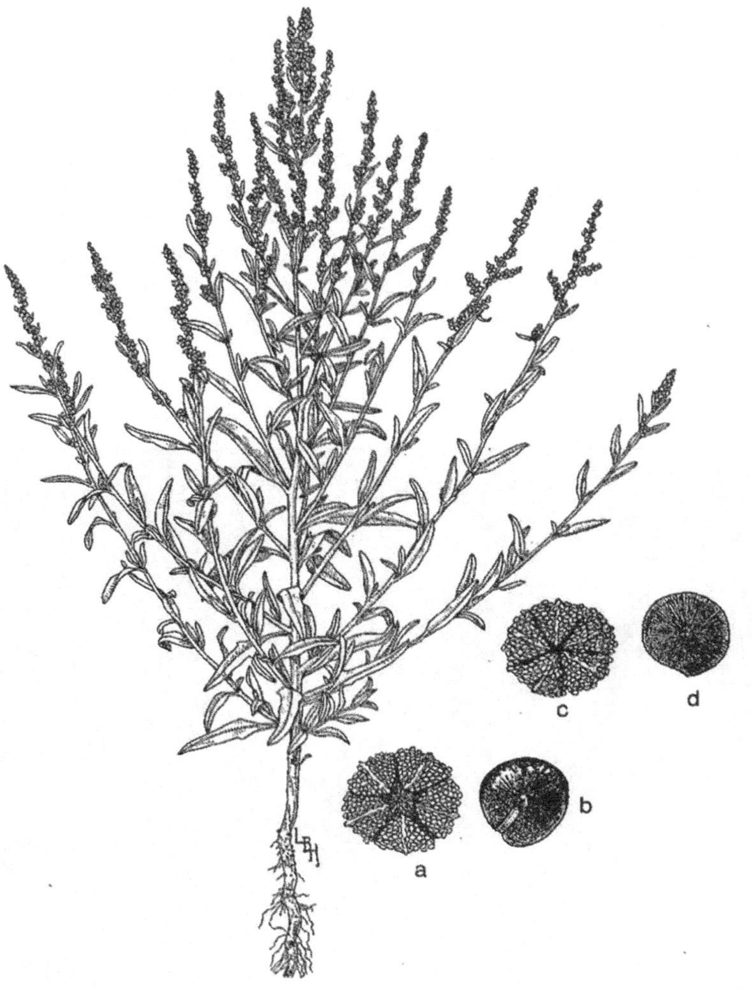

Fig. 9.5 Narrowleaf Goosefoot (*Chenopodium desiccatum*)—AMARANTHACEAE. Annual 1–4 ft. high branching well above base and reproducing only from seed. **Leaves**: Narrow, pointed, white beneath, green above, 13–50 mm long, and 2–8 mm wide (Fig. 9.5). **Flowers**: Stalkless, covered by white mealy scales, crowded on short branches in upper leaf axils and in *long leafless flowering branches at the tops of stems*. **Seeds**: Black shiny disks about 1 mm in diameter. **a**. Fruiting calyx. **b**. Seed Slimleaf Goosefoot (*Chenopodium leptophyllum*) is almost identical but has narrower leaves and dull seeds. **c**. Fruiting calyx. **d**. Seed (Fig. 9.5 from *An Illustrated Guide to Desert Weeds*, by Kittie F. Parker. © 1972 The Arizona Board of Regents. Reprinted by permission of the University of Arizona Press)

Narrowleaf Goosefoot (*Chenopodium desiccatum*) [A|EMN]

Narrowleaf Goosefoot expanded here at CWF last summer. It was surely present before 2017, but I had not noticed it. In 2017, it spread over about 2000 square feet in front of the tractor shed. It is shorter and less troublesome than Kochia, which had dominated the site.

The local deer squad and various insects eat Goosefoot, showing how native consumers might limit the abundance of native weeds (Fig. 3.1, Figs. 9.6 and 9.7).

Fig. 9.6 This photo shows how delicious native insects find Goosefoot leaves (photo © GR)

Fig. 9.7 Nettleleaf Goosefoot (*Chenopodiastrum murale*)—AMARANTHACEAE. A bushy annual 1–3 ft. high reproducing by seeds. Kittie Parker said it stinks (1972, p. 104), but others say it is odorless. I have not found one to try. **Leaves**: Dark green, triangular, eggshaped alternate, pointed tips with one to eight teeth on each side. **Flowers**: With a fine granular coating cluster on short branches in leaf axils. **Seeds**: Tiny, black 1–2 mm wide with a thin fruit wall and a distinct rim. **a**. Fruiting calyx. **b**. Achene with enclosed seed (Fig. 9.7 from *An Illustrated Guide to Desert Weeds*, by Kittie F. Parker. © 1972 The Arizona Board of Regents. Reprinted by permission of the University of Arizona Press)

Nettleleaf Goosefoot (*Chenopodiastrum murale*) [A|EFM]

I have not seen this weedy shrub (Fig. 9.8) in CWF yet, but it is probably here somewhere. Another common name for the plant is Australian Spinach though it is thought to have originated from the Mediterranean region. The CABI Invasive Species Compendium (CABI 2019) reports the plant is present in 43 countries.

Fig. 9.8 Nettleleaf Goosefoot (photo © Stefan.lefnaer)

Fig. 9.9 Monolepis (*Monolepis nuttalliana*)—AMARANTHACEAE. Annual 6–30 in. wide, 8 in. tall. Resembles other succulent species, but its *pointed and lobed leaves are unique*. **Leaves**: Bright green, alternate, 1–7 cm long, lanceshaped with lobes near base (not stipules). **Flowers**: Inconspicuous green flowers cluster in axils of leaves in late spring to midsummer. **Seeds**: Round, flat, 1 mm wide. **a**. Fruit. **b**. Seed (Fig. 9.9 from *An Illustrated Guide to Desert Weeds*, by Kittie F. Parker. © 1972 The Arizona Board of Regents. Reprinted by permission of the University of Arizona Press)

Monolepis (*Monolepis nuttalliana*) [A|EMN]

Monolepis (Fig. 9.10) prospers on sunny, dry alkaline soils, but it can grow in partial shade and moist soils. It is present in Alaska, the Lower 48, and Mexico. The entire plant is edible by wildlife and people and has medicinal uses. Seeds are included in the seed vaults (Genesys 2019).

Fig. 9.10 Monolepis (photo © Matt Lavin)

Fig. 9.11 Fivehook Bassia (Smotherweed) (*Bassia hyssopifolia*)—AMARANTHACEAE. Erect annual 2–5 ft. tall with stout branches from the main stem. **Leaves**: Alternate, narrow, pointed, 6–13 mm long. Each leaf axil has a cluster of shorter leaves or a short spike crowded with small leaves and flowers, all woolly with yellowish hairs. **Flowers**: Flower spikes longer at tips of stems sometimes forming branches 2–12 in. long. The tiny flowers have no petals. **Seeds**: Thin yellow hairy fruit has five separate chambers enclosing seeds. Each chamber develops a yellow hooked spine about 2 mm long. Seeds are oval, about 2 mm long with charcoal gray edges. **a**. Enlarged fruiting branch. **b**. Fruiting calyx. **b**. Seed (Fig. 9.11 from *An Illustrated Guide to Desert Weeds*, by Kittie F. Parker. © 1972 The Arizona Board of Regents. Reprinted by permission of the University of Arizona Press)

Fivehook Bassia (Smotherweed) (*Bassia hyssopifolia*) [P|I]

This weed's common name advertises the plant's principal means of dispersal: The hooks latch onto fabric and animal fur and ride along until brushed off or the fur sheds.

One ability of Fivehook Bassia (Fig. 9.12) botanists find most interesting is its occasional hybridization with Summer Cypress (*Kochia scoparia*). This occurs in North America but never in the plant's Eurasian home territory.

Fig. 9.12 Fivehook Bassia bushy growth (photo © Forest & Kim Starr)

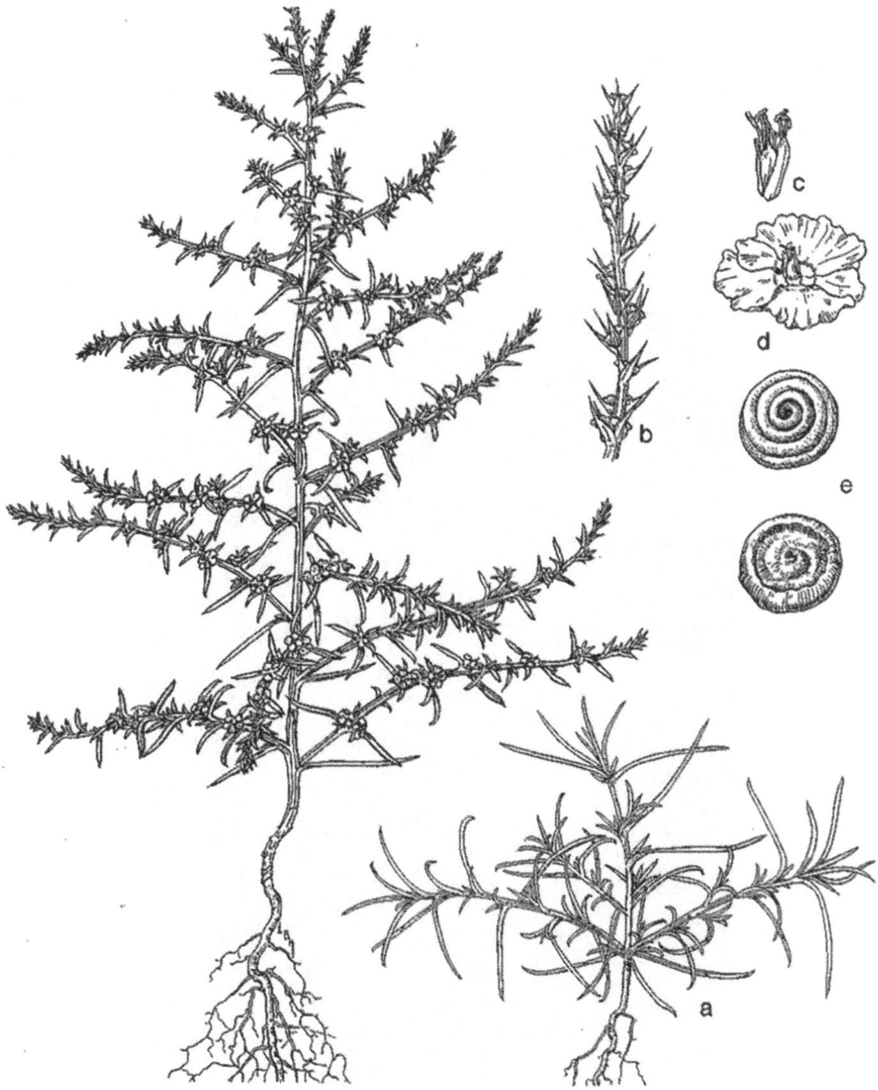

Fig. 9.13 Russian Thistle (*Salsola kali*)—AMARANTHACEAE. At maturity, a densely branched globular annual 6–72 in. tall with ridged, sometimes red, and sometimes red-streaked stems. Breaks from root when dry becoming a seed-dispersing tumbleweed. **Leaves**: Stiff awlshaped leaves ending in a spine replace soft early leaves. **Flowers**: Tiny, white, cluster at bases of leaves on upper branches. No petals. **Fruits**: Flower parts form wings that merge to cover the red fruit. Each fruit contains one gray to tan seed with the shape of a coiled embryo visible. **a**. Seedling. **b**. Fruiting branch. **c**. Flower. **d**. Fruiting calyx. **e**. Seeds (Fig. 9.13 from *An Illustrated Guide to Desert Weeds*, by Kittie F. Parker. © 1972 The Arizona Board of Regents. Reprinted by permission of the University of Arizona Press)

Russian Thistle (*Salsola Kali*) [A|EI]

People imported this prickly tumbleweed (Figs. 9.14) to the USA by accident in grass seed from Asia. Over the past century, the plant has spread across the arid West. For decades, it has been familiar on roadsides, disturbed areas, and overgrazed ranges everywhere. Dense tumbleweed drifts form against fences and other barriers and in one instance, the heap accumulating at the head of a train blocked the engineer's vision forcing the train to stop.

Tumbleweeds protect and restore soil better than some weeds. New shoots are edible, and some birds have learned to use the plants for cover. However, the weed is a misfortune for the West because it burns so well. When combined with other fuelweeds, it steps up fire frequency and this prevents succession back to the original vegetation.

Fig. 9.14 Russian Thistle (photo © Dcrjsr)

Fig. 9.15 Redroot Pigweed (*Amaranthus retroflexus*)—AMARANTHACEAE. Annual 1–3 m tall. Thick main stem red near base breaks at maturity to allow plants to tumble along dispersing seeds. **Leaves**: Lower leaves oval or lanceolate, hairy beneath at least along veins, 5–15 cm long. **Flowers**: Staminate (male) at tips of inflorescences with pistillate (female) flowers clustered beneath. Short spikes form in leaf axils, terminal spike longest, 5–20 cm. **Seeds**: Black to dark brown in a capsule less than 2 mm long with a lid that opens to reveal the tiny seed (Fig. 9.15 painting by Norman Criddle)

Redroot Pigweed (*Amaranthus Retroflexus*) [A|EMN]

Redroot Pigweed (Fig. 9.16) can grow to 3 m tall, but I have seen none more than half that. Size does not matter; however, the weed is so variable it can make a little tuft of flowers and seeds when it is 3 in. tall. I listed the plant as native, but it could have followed people into the Southwest from its tropical base farther south. Around CWF, it grows along roads, beside the Agua Fria River, and in gardens.

People cultivate Amaranthaceae for their seeds. Cultivars grow well in CWF with a modest amount of added water. The ones I grew for local birds produced heavy seed heads 30 cm long but the birds were not interested and the seeds fell to the ground. I did not follow their fate, but local harvester ants and rodents gathered some.

Fig. 9.16 Redroot Pigweed seed head (photo © Bogdan)

Fig. 9.17 Prostrate Pigweed (*Amaranthus blitoides*)—AMARANTHACEAE. Branching, hairless, prostrate annual to 1 m wide. **Leaves**: Alternate, spatulate blade 1–2 cm long with smooth edges and with stalks half as long as blade. Prominent lateral veins distinguish the species from Common Purslane. **Flowers**: Green 2–9 mm long pistillate flowers with three stigmas mixed with staminate flowers with three stamens. **Fruits**: Ovid utricle, the thin membrane covering the seed is 2 mm long, seeds black 1–2 mm in diameter. **a**. Spreading plant. **b**. Flowering branch. **c**. Rootstock. **d**. Utricle. **e**. Seeds (Fig. 9.17 drawing by Regina O. Hughes, USDA)

Prostrate Pigweed (*Amaranthus blitoides*) [A|EFN]

This native weed is unlikely to cause problems in CWF lawns and gardens. Though it has spread west with human roads and settlements, it is not very aggressive and it is not invasive. Where livestock grazing overwhelms native vegetation, however, Prostrate Pigweed joins other weeds colonizing the damaged areas.

Many weeds, like Prostrate Pigweed (Fig. 9.18), are edible and can entice livestock to remain on a site long after soil microorganisms and longer-lived plants have been trampled into dust. This alone can create a semi-permanent weedland. Because many weeds are short-lived and dry out during summer, fire frequency increases and locks the weedland in place.

Fig. 9.18 Prostrate Pigweed leaves and flowers (photo © SB Johnny)

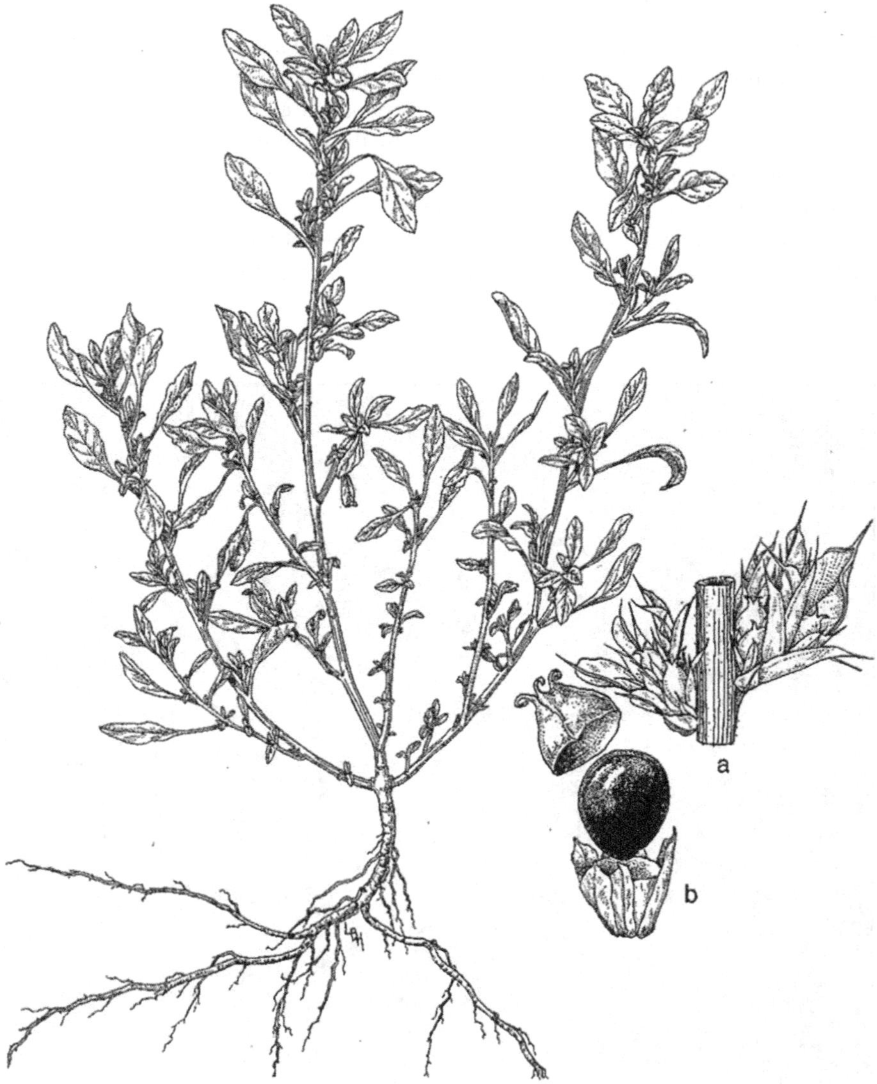

Fig. 9.19 Tumble Pigweed (*Amaranthus albus*)—AMARANTHACEAE. Branched sticky annual 6–48n tall with pale green or white stem, reproduces by seed. Often breaks off at base and blows along like a tumbleweed. **Leaves**: Alternate, 1–3 in. long, oblong, often purple beneath with veins and margins white edged. **Flowers**: Green in narrow clusters in leaf axils produces papery utricle that separates on an equatorial line freeing the top like a lid. **Seeds**: Shiny brown or black disks about 1 mm across. **a**. Stem with flower clusters in leaf axil. **b**. Seed emerging from papery fruit (Fig. 9.19 from *An Illustrated Guide to Desert Weeds*, by Kittie F. Parker. © 1972 The Arizona Board of Regents. Reprinted by permission of the University of Arizona Press)

Tumble Pigweed (*Amaranthus albus*) [A|N]

This small native tumbleweed is common in CWF, but less so than Russian Thistle the introduced tumbler. Many wildlife species have found uses for Tumble Pigweed (Fig. 9.20); the foliage hosts caterpillars of the butterfly *Pholisora catullus* (Common Sootywing) and at least three moth species. Mourning Dove and several songbird species eat the seeds. The plant is not desirable outside CWF, however because of its tendency to combine with introduced fuelweeds to create a fire hazard.

Fig. 9.20 Fruit showing utricle "lid" above the seed (photo © Fischer et al.)

Fig. 9.21 Palmer Pigweed (*Amaranthus palmeri*)—AMARANTHACEAE. Annual 1–6 ft. and sometimes 15 ft. tall. Thick main stem often turns red with age. **Leaves**: Lanceolate or eggshaped, alternate, hairless, 5–20 cm long, and 1–12 cm wide with prominent white veins on lower surface (**a**). **Flowers**: Staminate (male, **b**) and Pistillate (female, **c**) flowers on different plants, on long leafless spikes. Central spike 6–18 in. long. Both male and female flower parts stiffen to spines at maturity. **Seeds**: Dark brown, oval, about 2 mm long (**d**) (Fig. 9.21 from *An Illustrated Guide to Desert Weeds*, by Kittie F. Parker. © 1972 The Arizona Board of Regents. Reprinted by permission of the University of Arizona Press)

Palmer Pigweed (*Amaranthus palmeri*) [A|EMNP]

Amaranths are a minor irritant in crops and gardens and on the margins of lawns, roads, and other disturbed areas. However, Palmer Pigweed and other Amaranth species have many uses. There was one warning by Moore (2003) worth noting: "Do not grow where inorganic nitrates are added to the soil—or even overly rich [say from manure additions] soil; the Amaranths concentrate nitrates and can be accumulatively toxic" (Moore 2003, p. 29).

Palmer Pigweed (Fig. 9.22) and other native amaranths are consumed by numerous wildlife species (Fig. 3.1).

As many other weeds have done, Palmer Pigweed has evolved resistance to glyphosate (the main poison constituent of Round-Up).

Fig. 9.22 Palmer Pigweed (photo © GR)

Fig. 9.23 Kochia (Summer Cypress) (*Bassia scoparia*)—AMARANTHACEAE. Thornless bushy annual grows to 4 m tall. **Leaves**: Alternate, stalkless, more or less hairy, 2–5 cm long, 3–7 mm wide. **Flowers**: Solitary or paired in axils of leaf-like bracts, stalkless, green, with five triangular sepals and no petals. Grow on hairy spikes 5–10 cm long. **Seeds**: Persistent 2–3 mm wide sepals enclose one 1–2 mm wide, wedgeshaped, flat seed (Fig. 9.23 photo © GR)

Kochia (Summer Cypress) (*Bassia scoparia*) [A|IP]

Kochia originated in Europe and has become abundant throughout North American deserts. The plants produce seed when as small as 3 in. tall, and many more seeds up to their maximum height of about 13 ft. Large plants get so dense that they impede movement (Fig. 9.23). When they dry out in the fall, they are a serious fire hazard (Fig. 9.24).

The plant was present in a few spots around CWF until 2012 when it made rapid gains. Two years earlier the last of our rescued cows died. I stopped irrigating the pastures, and as the grass dried, Kochia and other weeds spread. Kochia has taken over with 2–3 ft. tall plants in the dry sections of the pastures and taller plants in the wetter areas. It does not grow in the shade of any of the trees.

Fig. 9.24 Dry Kochia stunted by drought (photo © GR)

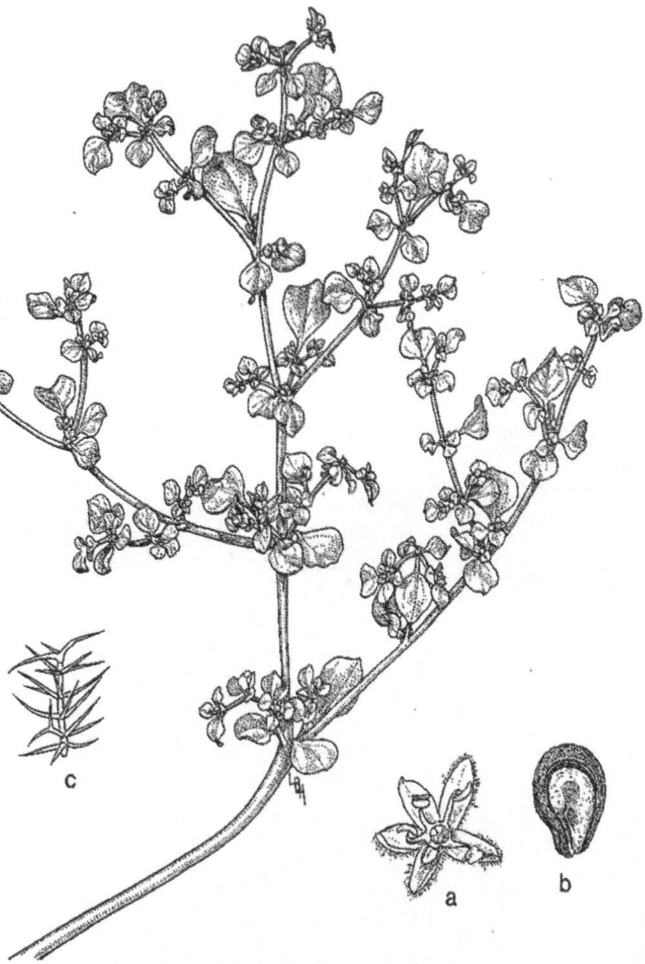

Fig. 9.25 Woolly Tidestromia (*Tidestromia lanuginosa*)—AMARANTHACEAE. Erect or prostrate annual appears after start of summer rains, 3–18 in. stems can reach 30 in.; reproduces from seed; white stems often bright scarlet by fall. **Leaves**: Hairy, opposite, oval, 8–50 mm long. The hair wears off the upper surface revealing the green beneath. The two opposite leaves are usually not the same size. **Flowers**: Yellow about 3 mm long in clusters in leaf axils; five stamens and five segments. **Seeds**: small, almost globular. **a**. Flower. **b**. Seed. **c**. Branched hair (Fig. 9.25 from *An Illustrated Guide to Desert Weeds*, by Kittie F. Parker. © 1972 The Arizona Board of Regents. Reprinted by permission of the University of Arizona Press)

Woolly Tidestromia (*Tidestromia Lanuginosa*) [A|N]

This native weed is abundant in North America from the southern Great Plains west to the Mojave Desert. It is uncommon in the cold Great Basin Desert. Parker (1972) says it is most conspicuous after summer rains when it forms whitish mats between desert shrubs and trees on mesas, slopes, and plains (Fig. 9.26). I have had limited experience with Woolly Tidestromia, but I expect hoeing and pulling will keep it in check.

Fig. 9.26 Woolly Tidestromia (photo © Pompilid)

9.2 Dogbane Family: APOCYNACEAE (Fig. 9.27)

Fig. 9.27 Periwinkle [*Vinca minor* (above) and *Vinca major* (below)]—APOCYNACEAE. Glossy green vines. **Leaves:** Opposite, *V. major* leaves are widest below the middle and have fine bristly hairs along the edges. *V. minor* leaves are widest at the middle. **Flowers**: Blue or violet with asymmetrical petals seeming to spin clockwise about the central funnel. **Fruits**: Two capsule-like pods (follicles). *V. majors* contain many seeds and *V. minors* contain 4–8. Other than seed number and size (Fig. 9.28), the species are very similar (Fig. 9.27 photo © GR)

Periwinkle (*Vinca major* and *Vinca minor*) [P|IM]

These evergreen perennials (Fig. 9.28) originated in the Mediterranean region and have been cultivated in Europe for centuries. They are common throughout North America and less common in South America. Though most nurseries sell these plants, they are invasive and are a threat to native plants in moist shady areas along desert streams and in moist canyons. They do not spread in dry uplands without supplemental moisture.

Fig. 9.28 Periwinkle leaves: *Vinca minor* above, *Vinca major* below (photographer unknown)

9.3 Milkweed Family: ASCLEPIADACEAE

Fig. 9.29 Inmortál (*Asclepias asperula*)—APOCYNACEAE. Unbranched stems ascending or lying on the ground. **Leaves**: Lanceolate 5–19 cm long, 6–27 mm wide, irregularly spaced, alternate. **Flowers**: umbels (flowers on stalks arising from a common point like an umbrella blown inside out) at ends of branches, 4–7 cm wide, peduncles 20–30 cm long; corolla green, bowl shaped with curved-ascending lobes 9–12 mm long; sessile hoods attach along the entire height of the column. **Fruit**: seedpods 3 in. long contain many flat seeds. Each seed has a tuft of fine 1 in. long, silky white hair (Fig. 9.29 photo © Max Licher)

Inmortál (*Asclepias Asperula*) **[P|NM]**

Widespread, but not abundant, Inmortál (Figs. 9.29 and 9.30), like other milkweeds, serves as a host for Monarch butterfly caterpillars. It has potent medicinal properties, but you should read Moore (2003) before trying it.

Fig. 9.30 Inmortál seedpods (photo © Stan Shebs)

Fig. 9.31 Broadleaf Milkweed (*Asclepias latifolia*)—APOCYNACEAE. Erect stems 40–110 cm tall. **Leaves**: Opposite, broad, elliptical, 5–14 cm long, 4–12 cm wide, short woolly when young. **Flowers**: Fragrant, green to cream-colored. Umbels at most upper nodes, 4–8 cm broad, sessile or with peduncles 2–3 cm long. Green calyx 4 mm long; corolla green with lobes 7–10 mm long, whitish to yellow-brown hoods. **Fruit**: Seedpods 7–9 cm long. **a.** Flower with reflexed petals with rising hoods and inward-pointing horns (Fig. 9.31 from *An Illustrated Guide to Desert Weeds*, by Kittie F. Parker. © 1972 The Arizona Board of Regents. Reprinted by permission of the University of Arizona Press)

Broadleaf Milkweed (*Asclepias latifolia*) [P|DN]

This big-leafed milkweed (Fig. 9.32) ranges from the southern Great Plains across the Chihuahuan and Sonoran Deserts to the Mojave Desert in the west and the southeastern Great Basin Desert in the northeast. Like all milkweeds, it has bitter sap and hosts Monarch Butterfly caterpillars.

Fig. 9.32 Broadleaf Milkweed (photo © Anthony Zukoff)

Fig. 9.33 Western Whorled (Horsetail) Milkweed (*Asclepias subverticillata*)—ASCLEPIADACEAE. Bushy perennial to 4 ft. tall reproduces by seed and from roots. **Leaves**: Two-four long narrow leaves 3–5 in. long to 8 mm wide in whorls at stem joints. **Flowers**: Greenwashed white in umbels at tips of branches and in leaf axils. **Seeds**: Copious, brown 6 mm long with tuft of hair in slender pointed pods 2–4 in. long. **a**. Seedpod. **b**. Flower. **c**. Seed with tuft of silky hair (Fig. 9.33 from *An Illustrated Guide to Desert Weeds*, by Kittie F. Parker. © 1972 The Arizona Board of Regents. Reprinted by permission of the University of Arizona Press)

Western Whorled (Horsetail) Milkweed (*Asclepias subverticillata*) [P|DN]

Similar in distribution to Broadleaf Milkweed (Figs. 9.31 and 9.32), Horsetail Milkweed (Fig. 9.34) is the most common milkweed around CWF. Quite similar species occur east (*A. verticillata*) and to the west (*A. fascicularis*), Horsetail Milkweed is preferred by Monarch Butterflies. Patches as small as six plants will attract Monarchs. Residents of the Western United States might wish to learn more and order milkweed seeds from the Southwest Monarch Study website (2019).

Fig. 9.34 Horsetail Milkweed (photo © GR)

Fig. 9.35 Climbing Milkweed (*Sarcostemma cynanchoides*)—APOCYNACEAE. Long twining stems 8–40 ft. long. **Leaves**: In pairs with slender stalks from 2 to 50 mm long. Width varies from 2 to 44 mm with tapered or rounded bases or both combined. Can be heartshaped, arrowshaped, or lobed 1–6 cm long. **Flowers**: 15–25 fragrant flowers on slender stalks in umbel clusters typical of milkweeds. Flowers white, ivory, or purple. **Seeds**: Seedpods brown, 4–9 cm long, and 8–16 mm wide. Brown seeds are 3–6 mm long with a tuft of soft hairs at narrow end (**a**. Leaf variation. **b**. Seed. **c**. Flower). (Fig. 9.35 from *An Illustrated Guide to Desert Weeds*, by Kittie F. Parker. © 1972 The Arizona Board of Regents. Reprinted by permission of the University of Arizona Press)

Climbing Milkweed (*Sarcostemma Cynanchoides*) [P|DN]

This native weed will climb on anything from fences to shrubs to trees. It spreads on dry wash banks, across gardens, and into cultivated fields. Flowers appear from May to September and seedpods persist until November.

Climbing Milkweed (Fig. 9.36) hosts Monarch and Queen Butterfly caterpillars but not as often as members of the milkweed genus *Asclepias*. There were many traditional uses for milkweeds and some of these continue today.

Fig. 9.36 Climbing Milkweed (photo © Curtis Clark)

9.4 Daisy Family: ASTERACEAE (Fig. 9.37)

Fig. 9.37 Rough Fleabane (*Erigeron divergens*)—ASTERACEAE. Perennial, small to about 1 ft. tall by 1 ft. wide, green, hairy, soft, stems trailing to ascending. First flowers about March 1. **Leaves:** 2–3 cm cm long, 2–3 mm wide; spatulate to linear, basal immature leaves often with 1–2 (rarely 4) pointed lobes; upper leaves partially clasping, often smaller than lower ones. **Flower** heads born alone 2–4 cm above last leaf; buds sometimes red. Hair on flower stalks longer and denser near and onto phyllaries. **Phyllaries** persistent, 40–50 in 2–3 series, lanceolate boat shaped with acute tips, villous. Involucre, shallow hemispheric cup 8–12 mm wide, becomes convex as it dries. **Ray flowers** 70–100 thin, longitudinally curled (some almost a tube), corollas lavender to white, 5–10 mm long, 0.6–1.2 mm wide; narrow, linear, pointed, boat shaped. **Disk flowers**: 150–200, yellow, tubular with five lobes, 10–20 silvery-white pappi shorter than corolla (Fig. 9.37 photo © GR)

Rough Fleabane (*Erigeron Divergens*) [P|DMN]

This small weed is common along roadsides and in most valley and foothill vegetation throughout arid North America and a few spots in South America.

The North American Ethnobotany Database (2019) lists several reports of the value of Rough Fleabane for good fortune, eyewash, childbirth, and more.

Around human settlements, Rough Fleabane (Fig. 9.38) colonizes exposed soil of roadsides, gardens, and construction sites. It has no barbs or thorns, and I have seen no cases where it invaded taller or denser vegetation. The standard controls—mulch, shade, and pulling—should work well. The plant's small flowers appear very early. They feed bees and other arthropods active on warm winter days and during early spring when few other flowers are available. The bugs the flowers attract supply food for insectivorous birds such as Shrikes. This is a desirable weed.

Fig. 9.38 Fleabane (photo © GR)

Fig. 9.39 Slimleaf Bursage (*Ambrosia confertiflora*)—ASTERACEAE. Bushy perennial 1–3 ft. tall reproducing by seed and by sprouts from slender creeping roots. **Leaves**: So hairy they appear gray; alternate, 2–5 in. long, with narrow lobes often divided into smaller lobes. **Flowers**: Spikes at tips of branches 2–5 in. long with tiny male flowers in green drooping involucre. Female flowers cluster in leaf axils below males; almost invisible until they mature into spiny little burs 2–3 mm long with 10–20 curved spines about 1 mm long with hooked tips. **Seeds**: Each bur encloses 1 or 2 achenes. **A**. Slimleaf Bursage showing finely divided leaves and male flowers above females in leaf axils. **Aa**. Enlarged male head with male flowers. **Ab**. Mature female flower (bur) with curved, hooked spines. **B**. Annual Bursage (*A. acanthicarpa*). Mature female flower head (bur) with straight spines (Fig. 9.39 from *An Illustrated Guide to Desert Weeds*, by Kittie F. Parker. © 1972 The Arizona Board of Regents. Reprinted by permission of the University of Arizona Press)

Slimleaf Bursage (*Ambrosia Confertiflora*) [P|ENT]

Not following my weed-control advice, I acted late and this weed formed a colony beside the CWF front gate. I have pulled the shoots every year for at least 10 years and still they come. The tenacity of this character is due to another negligence of mine; I forget to pull the shoots when they first appear in spring. The delay gives them time to harvest and store a little solar energy for next year. Slimleaf Bursage (Fig. 9.40) ranges from the southern Great Plains west to the Pacific coast and south to central Mexico.

Fig. 9.40 Slimleaf Bursage (photo © Max Licher)

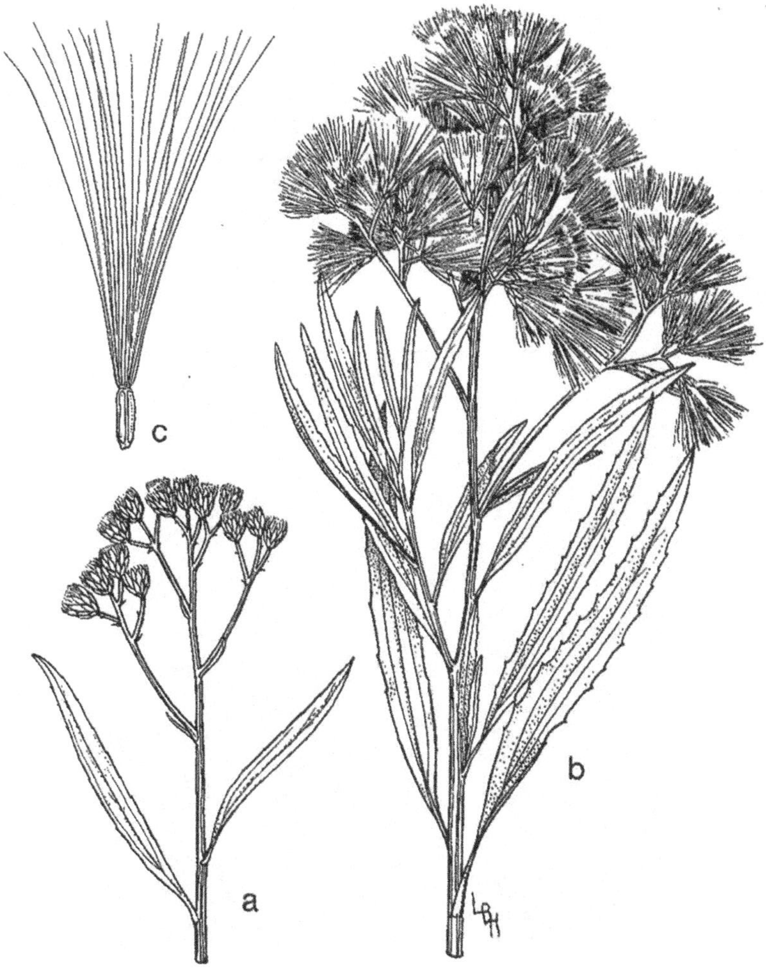

Fig. 9.41 Seepwillow Baccharis (*Baccharis salicifolia*)—ASTERACEAE. Shrubs with clusters of stems 1–4 m tall reproducing only by seed. Stems woody 12–25 mm in diameter at base. **Leaves**: Willowlike, 3–15 cm long, and 6–16 mm wide with small teeth. **Flowers**: Pale green to white, 5 mm wide in clusters at tips of main branches; male and female on separate plants. **Fruit**: Oblong achene 1–2 mm pale tan to greenish brown with 5 white ribs and a tuft of silky white hair 5 mm long. **a**. Branch from male plant with male heads. **b**. Branch from female plant with female flowers in fruit. **c**. Seedlike achene with tuft of silky hair (Fig. 9.41 from *An Illustrated Guide to Desert Weeds*, by Kittie F. Parker. © 1972 The Arizona Board of Regents. Reprinted by permission of the University of Arizona Press)

Seepwillow Baccharis (*Baccharis salicifolia*) [P|MN]

This big native shrub is common along streams and washes in arid southwestern North America, Central America, and South America. It is most abundant in the Arizona Uplands subdivision of the Sonoran Desert. It was common in the dry Salt River channel in Tempe. Livestock do not eat it so it increases where livestock grazing is heavy (Fig. 9.42).

Fig. 9.42 Seepwillow (photo © Krzysztof Ziarnek)

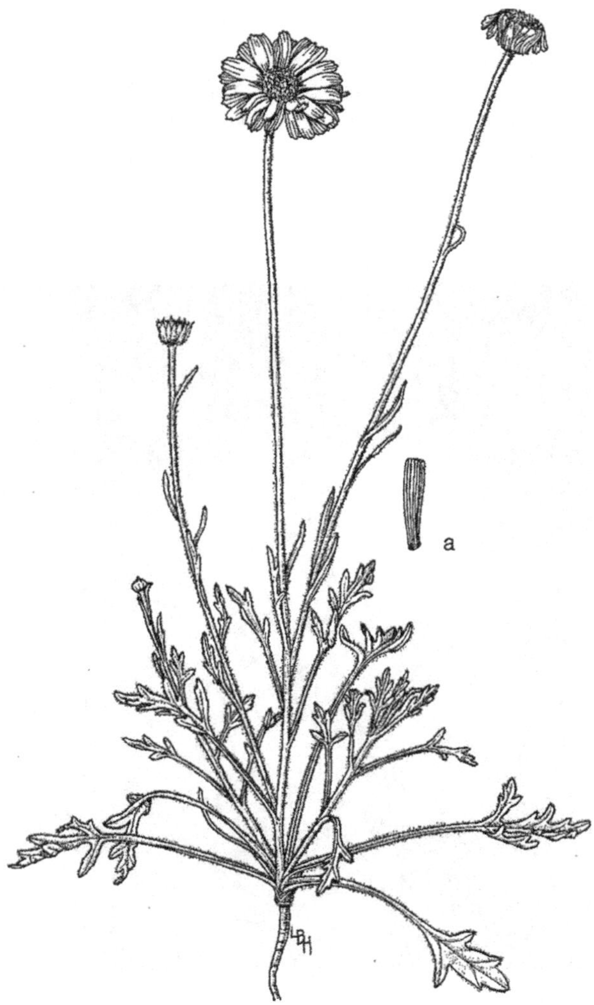

Fig. 9.43 Desert Marigold (*Baileya multiradiata*)—ASTERACEAE. Low, gray-green woolly biennial with a thick taproot and several woolly stems 6–18 in. tall. Stems have leaves on lower half and end with one flower head composed of many ray and disk flowers. **Leaves**: Alternate, woolly, lobed, and clustered at base of plant. **Flower heads**: Bright yellow, 2–4 cm wide with 20–50 ray flowers (petals). The rays dry and droop with age. Each head produces at least 100 rod-shaped disk flowers about 3 mm long, light brown with longitudinal ridges on the surface (Fig. 9.43 from *An Illustrated Guide to Desert Weeds*, by Kittie F. Parker. © 1972 The Arizona Board of Regents. Reprinted by permission of the University of Arizona Press)

Desert Marigold (*Baileya Multiradiata*) [B|DN]

This beautiful native of Western North America is common along roadsides below 5000 ft. in all four North American deserts. The flowers are the preferred caterpillar incubator for Desert-Marigold Moth (*Schinia miniana*), a flower moth of the Noctuidae family. I have not seen this moth, but it is present around the region (Fig. 9.44).

Fig. 9.44 Desert Marigold (photo © Chris English)

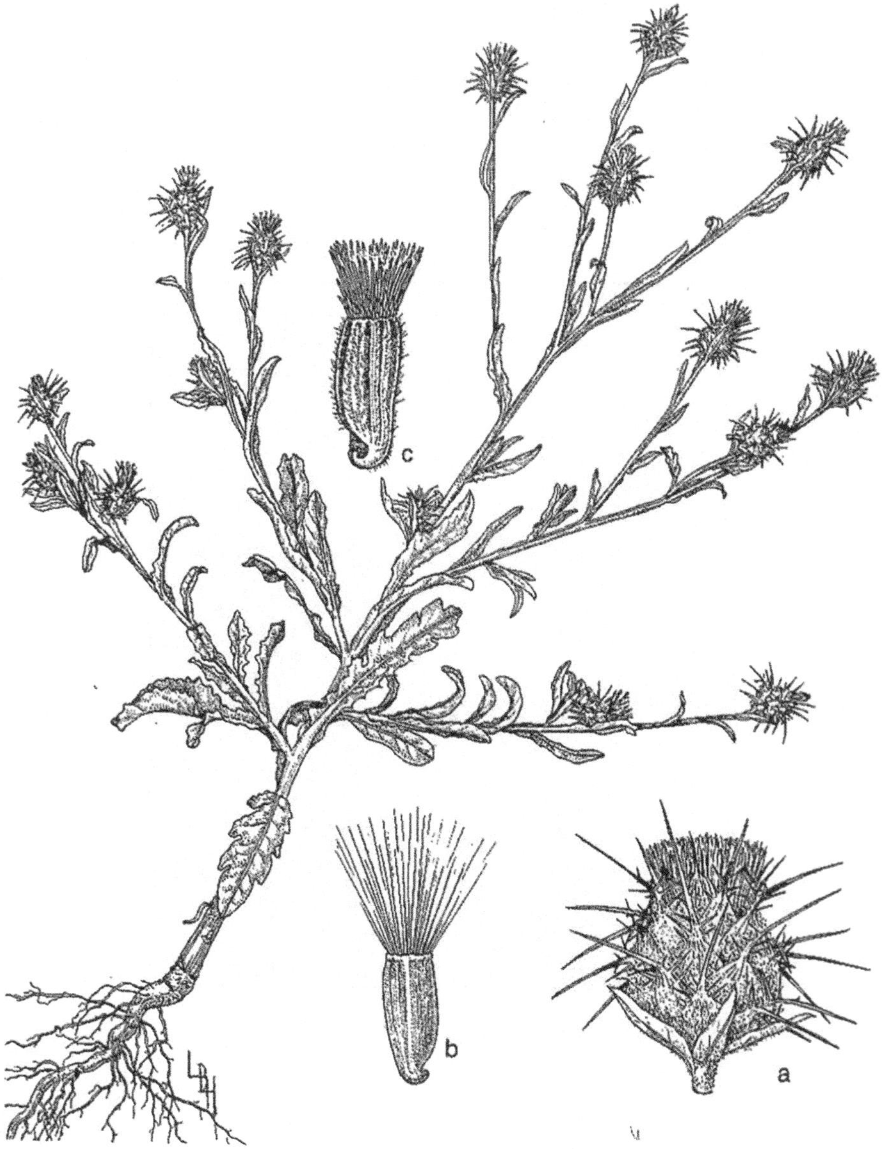

Fig. 9.45 Malta Starthistle (*Centaurea melitensis*)—ASTERACEAE. *Tomentose* grayish-green annual 1–2 ft. tall with winged stems. **Leaves**: Basal leaves 2–5 in. long lobed. Upper leaves alternate, narrow, and unlobed with bases extending down the stem as narrow wings. **Flowers**: Heads about 13 mm high with stiff bracts bearing slender spines 3–10 mm long incloses small tubular bright yellow flowers. **Seeds**: Achenes hairless, gray to brown, about 3 mm long with hooklike notch just above base. Has three rows of white bristles. **a**. Flower head enclosed in spiny bracts, spines of middle bracts branched near their base. **b**. and **c**. Achenes with hooklike notch (Fig. 9.45 from *An Illustrated Guide to Desert Weeds*, by Kittie F. Parker. © 1972 The Arizona Board of Regents. Reprinted by permission of the University of Arizona Press)

Malta Starthistle (*Centaurea Melitensis*) [A|P]

Malta Starthistle is present in all North American deserts and the east slope of the Andes Mountains in Argentina. People introduced it to the New World from the Mediterranean region of Europe and Africa. The first documented presence in the USA was at an adobe house built in San Fernando in 1797 (Encyclopedia of Life 2019). A single specimen, popped up in a Virginia Creeper bed at CWF last spring. The seeds are wind dispersed, so the weed will soon appear elsewhere in CWF. The sharp rather intimidating spines are more for defense from grazing than for seed dispersal (Fig. 9.46).

Fig. 9.46 Malta Starthistle (photo © Franco Folini)

Fig. 9.47 Russian Knapweed (*Acroptilon repens*)—ASTERACEAE. Perennial shrub 1–3 ft. high reproduces by seeds and shoots from underground rhizomes. Soft gray hair covers young stems and leaves. **Leaves**: Lower leaves 2–4 in. long, 12–36 mm wide with deep lobes. Upper leaves smaller and less lobed to lobeless. **Flowers** Cone-shaped flower heads 6–13 mm wide at branch tips contain 15–20, 5-lobed lilac to blue to rose pink flowers. Overlapping rows of pearly bracts surround the heads. Lower, outer bracts with rounded papery tips, upper, inner bracts with hairy tail-like tips. **Seeds**: Gray or ivory achenes 3–6 mm long have white bristles at the top. A. Five-lobed tubular flowers in heads surrounded by bracts. B. Achenes with remnant bristles (Fig. 9.47 from *An Illustrated Guide to Desert Weeds*, by Kittie F. Parker. © 1972 The Arizona Board of Regents. Reprinted by permission of the University of Arizona Press)

Russian Knapweed (*Acroptilon repens*) [P|ITZ-2]

This small Eurasian thistle has an attractive flower, but it is not a desirable plant. It forms dense stands from extensive root systems and it produces toxic chemicals. These competitive mechanisms give Russian Knapweed the power to invade and replace native vegetation in the Western United States. Once it becomes established, the root system makes eradication very difficult. This is one of those weeds where we follow the weed manager's rule, *Early Detection and Rapid Response* (EDRR) without question. Cut or pull the plant and repeat until no more sprouts appear (Fig. 9.48).

Fig. 9.48 Russian Knapweed (photo: U.S. Geological Survey)

Fig. 9.49 Rubber Rabbitbrush (*Ericameria nauseosa*)—ASTERACEAE. Woody shrub 18–72 in. high with many shreddy-barked flexible trunks from the base. Dense dull white to yellowish felt covers young branches. Somewhat sticky with resinous odoriferous gum. **Leaves**: Narrow, alternate, 1–3 in. long, less than 8 mm wide, and covered with the same felt-like hair. **Flowers**: Narrow, heads 6–13 mm long in rounded clusters at the ends of branches. Three to five rows of pointed bracts surround the 4–6 yellow flowers in each head. No outer ray flowers are present. **Seeds**: Slender achenes varying from smooth to tomentose and have pappus of dull white hairs. **a**. Flower head showing rows of bracts. **b**. Single flower. **c**. Stigmas (Fig. 9.49 from *An Illustrated Guide to Desert Weeds*, by Kittie F. Parker. © 1972 The Arizona Board of Regents. Reprinted by permission of the University of Arizona Press)

Rubber Rabbitbrush (*Ericameria nauseosa*) [P|DN]

Rabbitbrush is a large beautiful native shrub of the Western United States. People have spread it to Asia, South America, and a few Pacific islands, but nowhere does it flourish as it does in North American deserts. Its silvery leaves and brilliant yellow flowers brighten whole landscapes during its fall flower show.

Cows eat portions of Rubber Rabbitbrush after they have consumed all the grass and other more desirable plants. When this happens, other weeds increase. Fuelweeds such as Cheatgrass (*Bromus tectorum*) are often among the increasers; they support increased fire frequency exceeding even the colonizing ability of Rabbitbrush. After a few fires, only short-lived annual weeds remain.

This is a useful plant for restoring disturbed ground in arid environments. It is wildlife friendly, protects the soil, and requires little to no care (Fig. 9.50).

Fig. 9.50 Rubber Rabbitbrush (photo © Fred Bauder)

Fig. 9.51 Macoun's Rabbit-Tobacco (*Pseudognaphalium macounii*)—ASTERACEAE. Bushy annual or biennial to 90 cm tall with thin white distally tomentose stems. Flowers from July to October. **Leaves** alternate, 2–10 cm long by 3–13 mm wide. Tiny **flowers** numerous, yellow, in corymbs. **Seeds** in warty achenes (photo © GR)

Macoun's Rabbit-Tobacco (*Pseudognaphalium macounii*) [A-B|DN]

Rabbit-Tobacco grows in scattered stands in North America from Canada to Mexico and from the Pacific to the Atlantic across an elevational range of 150–8500 ft. I do not know if the plant has food or medicine values, and I have not seen rabbits smoke or chew it. The plant is rare at CWF and might not be as aggressive as other natives such as Palmer Pigweed. The one plant present in 2018 failed to reseed (Fig. 9.52).

Fig. 9.52 Macoun's Rabbit-Tobacco flowers (photo © Max Licher)

Fig. 9.53 Canada Thistle (*Cirsium arvense*)—ASTERACEAE. Plants 1–4 ft. high reproduce from seed and an extensive root system. Stems green and smooth. **Leaves**: Alternate, sessile, and clasping. Very spiny, lobed, to eight inches long and one inch wide. **Flowers:** 1–2 cm wide, contain many small rose-purple tubular flowers. Bracts enclosing flower heads are spineless. Male and female flowers appear on separate plants (dioecious). **Seeds**: Bone or tan colored achenes 3 mm long, oblong, smooth, with a tuft of feathery hairs remaining until maturity. **a**. Underground roots. **b**. Flower head. **c**. Achene with tuft of hair (Pappus). **d**. Mature achene without tuft of hair, shows blunt apex with central tubercle (Fig. 9.53 from *An Illustrated Guide to Desert Weeds*, by Kittie F. Parker. © 1972 The Arizona Board of Regents. Reprinted by permission of the University of Arizona Press)

Canada Thistle (*Cirsium arvense*) [P|EIMT]

This Eurasian perennial is one of the most dreaded weeds in New World drylands. It grows in crops, fields, livestock ranges, and other disturbed areas. From those disturbed sites, Canada Thistle spreads into and replaces natural vegetation. The first Arizona record is from Flagstaff in 1920. By 1970, colonies had appeared across Coconino and Yavapai Counties.

Though it is not a native, Canada Thistle is a food source for many insects and for Lesser Goldfinches. Parts of the plant are edible, and Wikipedia notes the Native American Cherokee used the feathery Pappus to fletch blowgun darts. These uses do not compensate for losing native vegetation to this alien invader.

Eradication of established colonies by mechanical means is difficult. I pulled aboveground shoots of a CWF colony for 10 years without success. New shoots appeared throughout the growing season making it difficult to pull them quick enough to prevent energy transfer to the roots (Fig. 9.54).

Fig. 9.54 Canada Thistle (photo © Põldohakas)

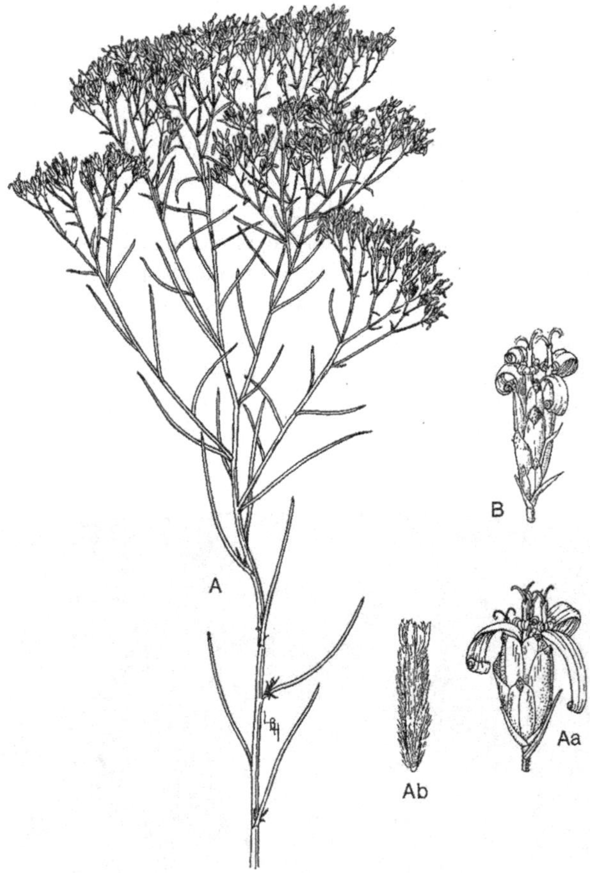

Fig. 9.55 Broom Snakeweed (*Gutierrezia sarothrae*)—ASTERACEAE. Small shrub 6–24 in. tall. Many thin branching stems flowering at their tips. **Leaves**: Each stem has many thin leaves 1–4 cm long. **Flowers**: At tips of branches, tiny bright yellow flowers appear summer through fall. **A**. Branches topped by clusters of tiny yellow flowers. **Aa**. Flower head 3–6 mm long with 3–8 central flowers and 3–8 marginal flowers. **Ab**. Achene 1 mm long with short Pappus. **B**. Narrower flower head of similar Threadleaf Snakeweed (*G. microcephala*) with 1–3 central flowers and 4–5 marginal flowers (Fig. 9.55 from *An Illustrated Guide to Desert Weeds*, by Kittie F. Parker. © 1972 The Arizona Board of Regents. Reprinted by permission of the University of Arizona Press)

Broom Snakeweed (*Gutierrezia sarothrae*) [P|MN]

This common native weed is one of our leading protectors of damaged land. It appears along roads, powerline and pipeline corridors, and on overgrazed livestock ranges. Where livestock grazing removes plants and breaks up soil surfaces, Broom Snakeweed will perform its duty and cover the ground. Cows and sheep dislike the plant and Ranchers sometimes blame it for loss of productivity of the range. Nonsense. Snakeweed protects soils from wind and water erosion. If ranchers remove the cattle, other plant species will replace snakeweed in the normal way of plant succession. However, alien weed species introduced by people often interrupt normal succession. Some of them promote fire, replace Snakeweed, and prevent the return of native plants (Fig. 9.56).

Fig. 9.56 Snakeweed (photo © Max Licher)

Fig. 9.57 Stinknet (*Oncosiphon piluliferum*)—ASTERACEAE. Small shrub with bright yellow flowers. South African annual with persistent roots. Small, less than 2 ft. tall. One to five or more thin stems arising from base, sparse alternate **leaves** with glandular hairs, striking yellow **flowers** in small tight balls less than 10 mm diameter. Fruit: Cylindrical smooth or ribbed achenes, resin-gland dotted face, less than 1 mm long. Stinky (Fig. 9.57 photo © Max Licher)

Stinknet (*Oncosiphon piluliferum*) [A|FIMT]

The first thought produced by Stinknet (also called Globe Chamomile) is its bright yellow flowers are beautiful. The next thought for most people, however, is something stinks. Stinknet produces resinous aromatic sap that smells like a rotten pineapple. The odor plus the tendency for the plants to grow in tight formation impedes outdoor activity. Even worse, Stinknet is a strong competitor that replaces native plants. But worse still, the dry plants are flammable and encourage destructive wildfires. Once Stinknet invades, the quality of natural habitats declines and many soil organisms, native plants, and native animals disappear.

Stinknet is spreading across the hot deserts of California and Arizona. Botanists have not collected the plant above 2300 ft. in Arizona, and I assumed at 4500 ft., CWF winters would be too cold for Stinknet. I did not even include it in the list of future weeds in *Weeds of Dewey-Humboldt, Arizona. Oops.*

However, in June 2019, I found one Stinknet growing beside the road about 200 m south of CWF (Fig. 9.58).

Fig. 9.58 Stinknet found 200 m southwest of CWF (photos © GR)

Fig. 9.59 Sunflower (*Helianthus annuus*)—ASTERACEAE. Annual 1–4 m tall reproducing only by seed. **Leaves**: Alternate, egg or heartshaped, toothed edges, pointed, 2–14 in. long, and 1–7 in. wide. **Flowers**: Yellow ray or marginal petals 2–5 in. wide surround many small female disk flowers that produce the achenes. **Seeds**: Achenes (shells) are 3–13 mm long and 3–8 mm wide, dark gray with black spots and pale stripes. **a**. The upper part showing flowers and leaves. **b**. Large basal leaf. **c**. Achene with dark spots and slightly hairy top (Fig. 9.59 from *An Illustrated Guide to Desert Weeds*, by Kittie F. Parker. © 1972 The Arizona Board of Regents. Reprinted by permission of the University of Arizona Press)

Sunflower (*Helianthus annuus*) [A|DEN]

Sunflowers are one of the North America's most productive and beautiful weeds. They are present but not invasive in other arid parts of the world. Essential for many wildlife species, the plant serves as host for at least three species of local butterflies: California Patch, Bordered Patch, and Painted Lady. Lesser Goldfinches and an occasional member of other species eat the leaves and seeds. Bees, butterflies, and other insects feed on the nectar and pollen.

Cultivators have produced many colors and sizes of sunflowers and they all grow well in CWF. Cultivars and wild plants often cross making cultivated varieties difficult to classify (Fig. 9.60).

Fig. 9.60 Sunflowers (photo © GR)

Fig. 9.61 Telegraph plant (*Heterotheca grandiflora*)—ASTERACEAE. Annual 1–2 m tall branching toward top like a telecommunications pole. **Leaves**: Hairy, basal rosette, smaller above. **Flowers**: Yellow 25–40 ray flowers 5–8 mm long surround 30–75 disk flowers with 4–6 mm corollas. **Fruit**: The disk and ray florets drop away to leave a spherical head of achenes (fruit), each with a long white pappus. Ray fruit 2–5 mm, disk fruit 4–6mm (Fig. 9.61 photo © Forest & Kim Starr)

Telegraph Plant (*Heterotheca grandiflora*) [A|N]

Common around CWF, and present in the Sonoran, Mojave, and Great Basin deserts, Telegraph plant is not a difficult plant to control using standard mechanical techniques. Its flowers attract butterflies and other insects, making it desirable for wildlife. However, the plant is too rough and shaggy to appeal to flower gardeners (Fig. 9.62).

Fig. 9.62 Telegraph plant (photo © GR)

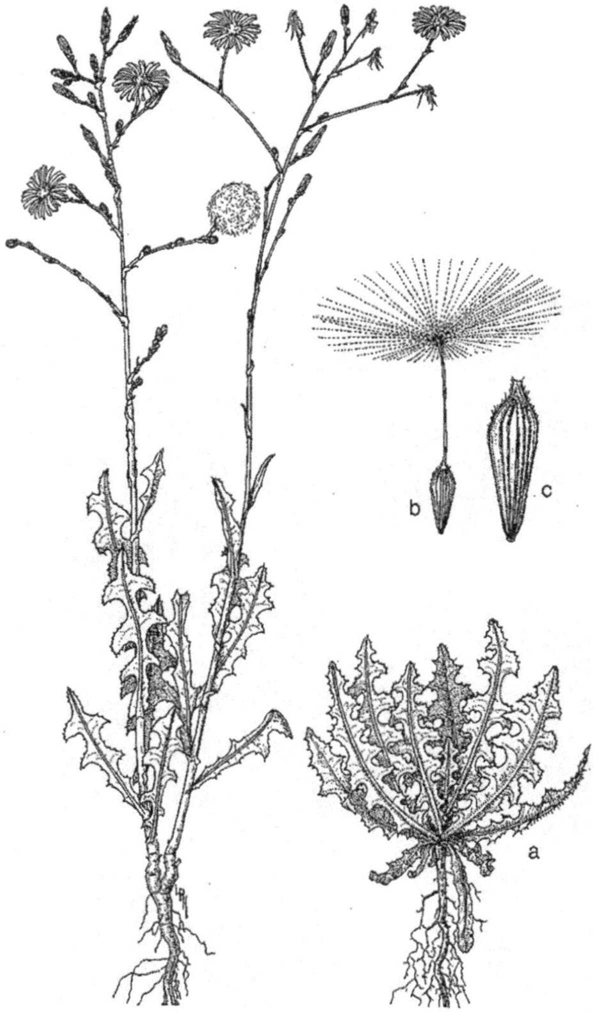

Fig. 9.63 Prickly Lettuce (*Lactuca serriola*)—ASTERACEAE. Annual or biennial 2–6 ft. tall with milky sap. **Leaves**: Alternate, blue-green, stalkless, clasp stem with 2 earlike lobes. Lower leaves 2–10 in. long lobed or unlobed with prickles along edges, the large white midvein, and other veins underneath. Upper leaves smaller and unlobed. **Flowers**: Many yellow heads 3–8 mm wide on short stalks and composed of ray flowers. **Fruit**: 6–30 flattened light gray eggshaped achenes about 3 mm long. **a**. Basal rosette of young plant. **b**. Achene with its parachute of hairs. **c**. Enlarged achene with linear ridges and short bristles near top (Fig. 9.63 from *An Illustrated Guide to Desert Weeds*, by Kittie F. Parker. © 1972 The Arizona Board of Regents. Reprinted by permission of the University of Arizona Press)

Prickly Lettuce (*Lactuca serriola*) [AB|MT]

This Eurasian annual is a cosmopolitan troublemaker familiar to every gardener worldwide. Though the plant's leaves are edible when young and tender, they cannot compare to Collards, Kale, or Spinach. The plant is a persistent pest during the summer, but a sharp hoe before the plumed seeds form will prevent a full-scale outbreak.

Numerous insect species spend time on Prickly Lettuce flowers and foliage. I have not found reference to any local bugs or butterflies that prefer it to other plants, but the possibility exists (Fig. 9.64).

Fig. 9.64 Prickly Lettuce (photo © Jeantosti)

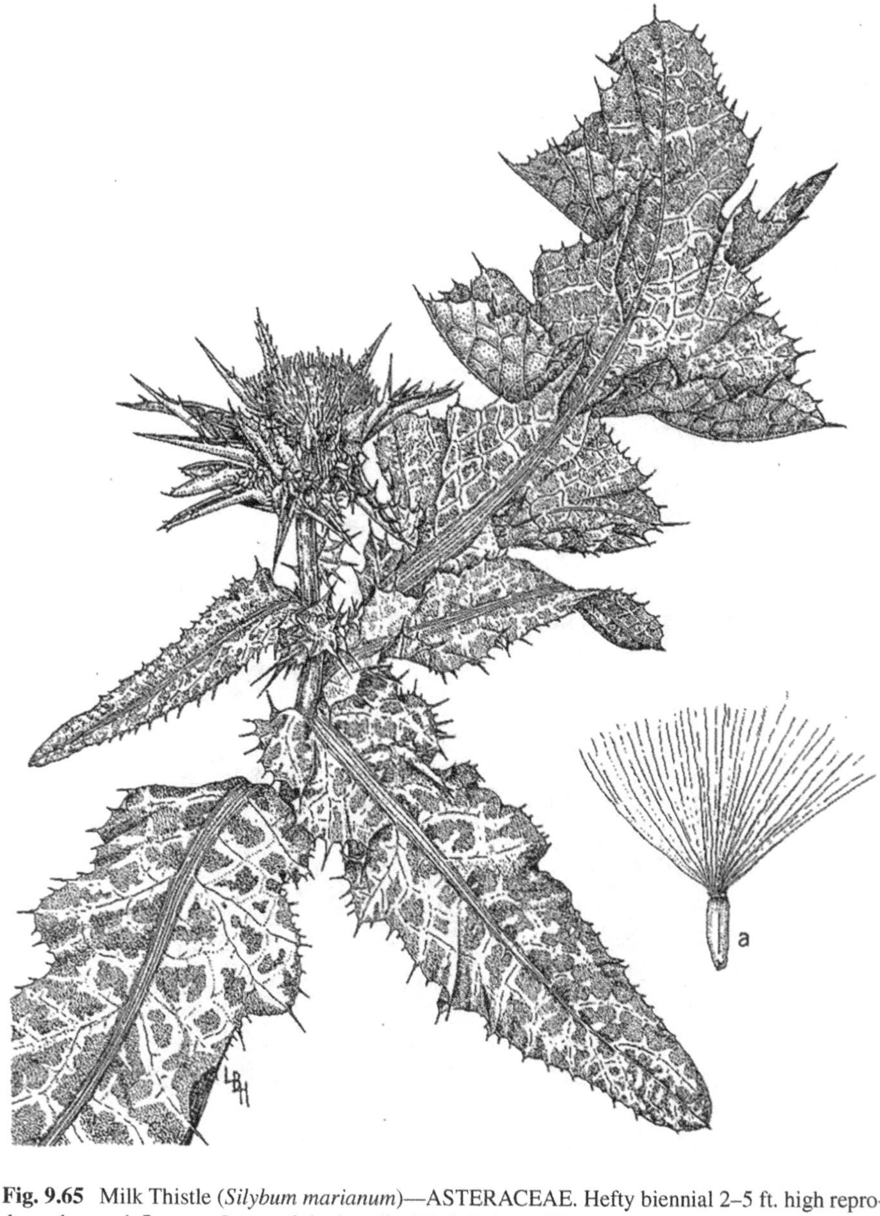

Fig. 9.65 Milk Thistle (*Silybum marianum*)—ASTERACEAE. Hefty biennial 2–5 ft. high reproduces by seed. **Leaves**: Lower, lobed, stalked 1–2 ft. long. Upper, progressively shorter and less lobed. Two earlike lobes clasp the stem. Edges with many short yellow spines, midvein with white lines, and surface with scattered white patches. **Flowers**: Dense heads of red-purple, tubular flowers with five long narrow lobes. Leathery bracts around each flower head in several overlapping tiers, spine-margined, tipped by stiff spines 1–4 cm long. **Seeds**: Achenes shiny, mottled buff and brown about 6 mm long. Each bears a tuft of white scaly bristles (**a**) (Fig. 9.65 from *An Illustrated Guide to Desert Weeds*, by Kittie F. Parker. © 1972 The Arizona Board of Regents. Reprinted by permission of the University of Arizona Press)

Milk Thistle (*Silybum marianum*) [B|EM]

Milk Thistle is a Mediterranean introduction rare around CWF. It is more common in the Arizona Upland subdivision of the Sonoran Desert below 3500 ft. Not restricted to drylands, the species appears in moist coastal areas and in the southeastern Great Plains of North America (Fig. 9.66).

Fig. 9.66 Milk Thistle (photo © Flagstaffotos)

Fig. 9.67 Bull Thistle (*Cirsium vulgare*)—ASTERACEAE. This big thorny biennial matures when 1–10 ft. tall. It forms a leaf rosette and a deep taproot the first year and a flowering stem the second. The stem has several spine-tipped wings along its full length. **Leaves**: spined margins, gray-green, and lobed. Rosette leaves are 6–10 inches long; upper leaves are smaller. **Flowers**: Inflorescence is 1–2 inches wide. **Seeds (c)** are about 1 cm long. (Painting in: *Flora Batava* vol. 4, (1822) Public Domain)

Bull Thistle (*Cirsium vulgare*) [B|EM]

This large Eurasian introduction is unpalatable for most livestock, but not rabbits and other small rodents including our local pocket gophers. The flowers are a favorite of bees, butterflies, and other insects. Goldfinches eat the seeds.

Bull Thistle is easy to eradicate if desired. It reproduces by dispersing seeds not by sprouting from roots. Use a sharp shovel to cut the root top below the basal leaves before seeds mature. The plants often recover, but one shoveling per year is usually enough to starve the roots until they die.

Young stems and roots are edible and there are medicinal uses (Fig. 9.68).

Fig. 9.68 Bull Thistle (photo: USDA)

Fig. 9.69 Spiny Sowthistle (*Sonchus asper*)—ASTERACEAE. Hollow stemmed annual 1–4 ft. tall. **Leaves**: Alternate with deep lobes and *soft prickle-toothed edges*. Lower leaves to 12 in. long, stalked with 5–11 lobes on each side. Mid and upper leaves not stalked, clasp the stem with large round lobes. Highest leaves not lobed. **Flowers**: Numerous 1–3 cm wide made up of yellow petal-like ray flowers. **Seeds**: Achenes reddish-brown, flat, ringed with a narrow wing, oval, 2–3 mm long with three ribs. Several achenes held together by the tangled hairs (6–10 mm long) of the pappus. **a**. Leaf with earlike clasping lobes. **b**. Flower. **c**. Achene (Fig. 9.69 from *An Illustrated Guide to Desert Weeds*, by Kittie F. Parker. © 1972 The Arizona Board of Regents. Reprinted by permission of the University of Arizona Press)

Spiny Sowthistle (*Sonchus asper*) [A|ET]

An Old World introduction to North and South America, Spiny Sowthistle is a tireless colonist of roadsides, lawns, gardens, fields, and other disturbed sites. The plants' hollow stems make them vulnerable to any form of mechanical control such as a swift kick to the base. Easy to remove, but difficult to eradicate, the plants spring from pappused (bristled) achenes spreading on spring breezes every year (Fig. 9.70).

Fig. 9.70 Spiny Sowthistle Dandelion-like flowers (photo © Matt Lavin)

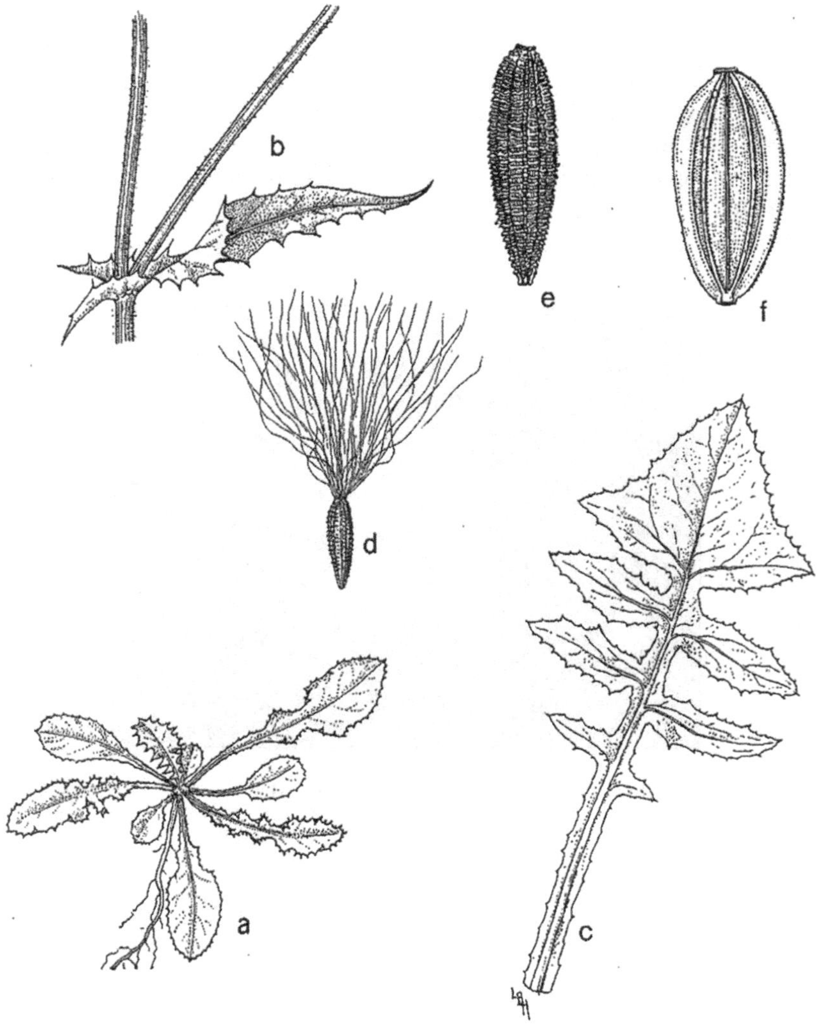

Fig. 9.71 Annual Sowthistle (*Sonchus oleraceus*)—ASTERACEAE. Single stem to 6 ft. tall. *Marginal prickles on leaves are smaller than those of Spiny Sowthistle above*. **a**. Main leaves of young plant showing large, one to three lobes, and a triangular-shaped tip lobe. **b**. Leaves clasp stem with two sharp projections. **c**. Stalked basal leaf. **d**. Achene with tuft of fine hair. **e**. Achene with five to seven indistinct ribs and strong cross wrinkling in furrows between ribs. **f**. *Achene of Spiny Sowthistle for comparison* (Fig. 9.71 from *An Illustrated Guide to Desert Weeds*, by Kittie F. Parker. © 1972 The Arizona Board of Regents. Reprinted by permission of the University of Arizona Press)

Annual Sowthistle (*Sonchus oleraceus*) [A|ET]

Annual Sowthistle resembles Spiny Sowthistle but its leaves do not look so dangerous. The toothy leaves of both plants are scratchy, but they are too thin to do much damage. The plant's stem is hollow leaving the whole business vulnerable to a swift kick.

A native of Europe and North Africa, Sowthistle is common at CWF often appearing on lawn and garden margins and poking up between other plants. Achenes with tufts of fluffy white hairs soon replace the spring and summer blooms. The plants are easily pulled or mashed when they appear in spring (Fig. 9.72).

Fresh Sowthistle leaves are bitter, but become mild when blanched or boiled. They contain desirable nutrients and high protein content.

Fig. 9.72 Annual Sowthistle (photo © Alvesgaspar)

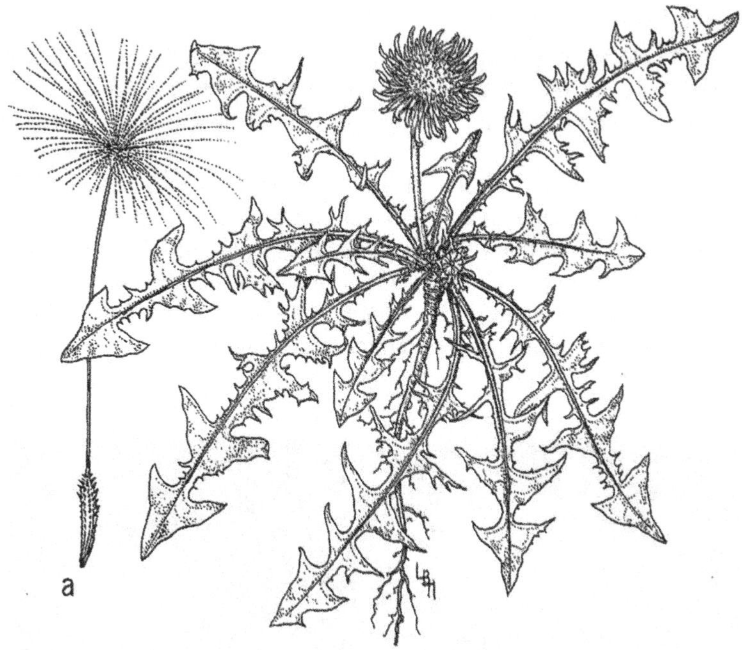

Fig. 9.73 Dandelion (*Taraxacum officinale*)—ASTERACEAE. A *short perennial* with deep roots that form new shoots on root crowns. **Leaves**: Clustered in a rosette at the base of the plant, sometimes upright. Variable length 2–12 in., lobed with pointed tips. **Flowers**: Solitary heads 1–2 in. wide composed of 40–100 ray flowers with five notches in the tip. **Seeds**: Achenes green or light brown, about 3 mm long, with 5–8 ribs on each side and minutely toothed with tiny curved spines on upper margins. The achene has a tuft of fine white persistent hairs. **a**. Achene with hair wind sail (Fig. 9.73 from *An Illustrated Guide to Desert Weeds*, by Kittie F. Parker. © 1972 The Arizona Board of Regents. Reprinted by permission of the University of Arizona Press)

Dandelion (*Taraxacum officinale*) [P|EM]

This may be our most familiar and beloved alien weeds. Found worldwide in temperate latitudes (not too dry, wet, or cold), Dandelion is from Europe, brought to the USA by early travelers. Watson (1871) reported the plant was growing in such isolated areas in the Great Basin (Utah–Nevada) he wondered if it might somehow be a native. Today, the consensus is people carried Dandelions with them from Europe to the New World, India, China, South Africa, Asia, and Australia. The plant's ability to disperse on the wind and establish in natural and human-caused disturbances let it spread across the new lands in a few years (Fig. 9.74).

Dandelion is difficult to eradicate without resorting to risky herbicides, but apart from a few crazed perfectionists, most of us are content to have a few make-a-wish puffs scattered across our lawns. Besides, this weed is a staple for many wildlife species and for humans. It goes into salads and stew pots, and we have all heard of Dandelion Wine. Moreover, the plant is probably the source of more botanical remedies for human ailments than any other species (Moore 2003).

Fig. 9.74 Ripe seed head (photo © GR)

Fig. 9.75 Crownbeard or Cowpen Daisy (*Verbesina encelioides* var. *exauriculata*)—ASTERACEAE. Grayish-green annual with a strong odor. Grows 1–4 ft. high. **Leaves**: Somewhat toothed, hairy beneath, and green above. Opposite lower leaves are narrow triangles 2–8 in. long including stalks. Upper leaves are alternate, shorter, often with a pair of leaf-like lobes at the stem. **Flowers**: Heads on long stalks 2–5 cm wide with 10–15 rays. Ray flowers are yellow, 8–13 mm long with three notches at the top. Disk flowers are yellow and tubular. **Achenes** flattened, gray-brown, about 6 mm long with broad wings along margins. Two short awns arise on each side of the achene tip. **a**. Achene (Fig. 9.75 from *An Illustrated Guide to Desert Weeds*, by Kittie F. Parker. © 1972 The Arizona Board of Regents. Reprinted by permission of the University of Arizona Press)

Crownbeard or Cowpen Daisy (*Verbesina encelioides* Var. *exauriculata*) [A|FMN]

This North American annual forms thickets along roads after summer rains. It joins Common Cocklebur as one of the few New World plants colonizing the Old World. The species name, *encelioides*, means resembling *Encelia farinosa* a similar yellow-flowered plant of the Sonoran Desert.

Crownbeard is allelopathic which means it produces biochemicals that block the germination and growth of other plant species. It can form dense stands with few other plants present. Crownbeard grows best in sunny locations and loses ground in the shade of larger plants.

Crownbeard attracts several species of flower moths. The caterpillars of at least one, the Alluring Schinia Moth (*Schinia siren*) feed exclusively on Crownbeard. I have not found evidence of this moth in CWF, but that might be because of the small number of Crownbeard (Fig. 9.76).

Fig. 9.76 Crownbeard (photo © GR)

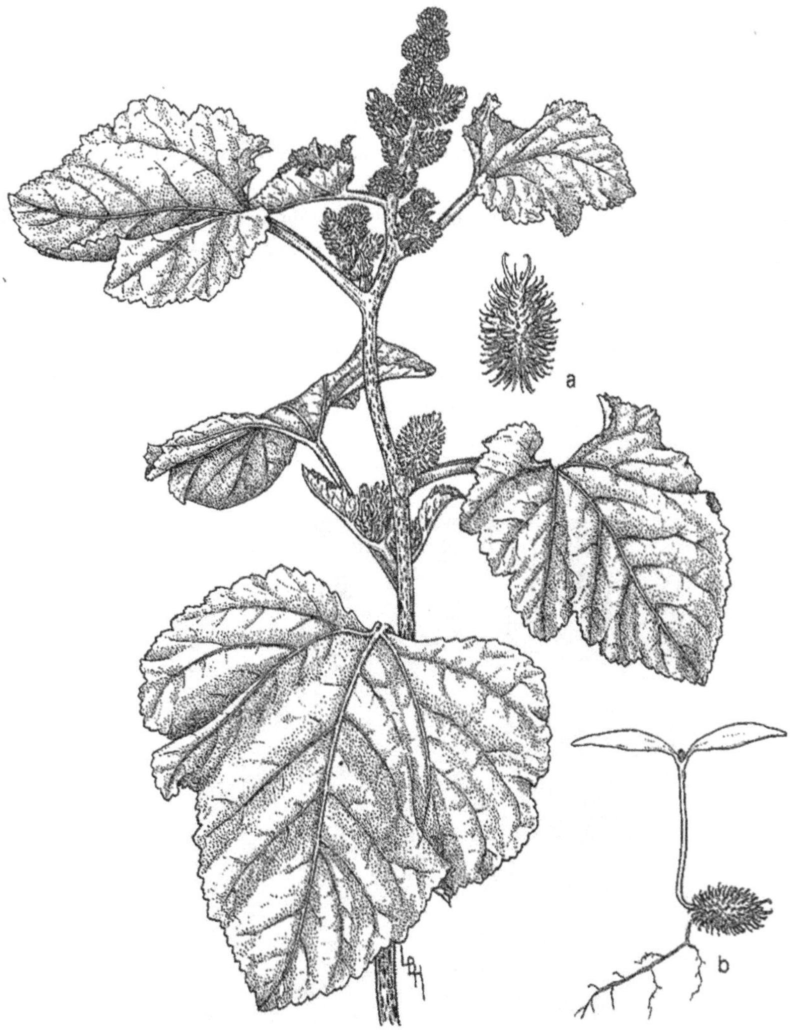

Fig. 9.77 Cocklebur (*Xanthium strumarium*)—ASTERACEAE. Two to three feet high. **Leaves**: Longstalked triangular somewhat lobed leaves 2–14 in. long and 1–8 in. wide. The leaves feel like sandpaper. **Flowers**: Inconspicuous male flowers in round clusters at the top, conspicuous female burs below. Burs, 12–25 cm long and covered by spines, enclose two female flowers. **Seeds**: Burs containing two blackish achenes are persistent. A pair of 13 mm long seeds in each achene remain fertile for many years. **a**. Spiny female bur. **b**. Seedling emerging from bur (Fig. 9.77 from *An Illustrated Guide to Desert Weeds*, by Kittie F. Parker. © 1972 The Arizona Board of Regents. Reprinted by permission of the University of Arizona Press)

Cocklebur (*Xanthium strumarium*) [A|MNT]

Native annual Cocklebur is one of the few New World weeds successful in the Old. It is present in Southern Europe, south central Asia, southern South America, and elsewhere. The plants colonize irrigated fields and moist disturbed sites in washes, floodplains, and pond margins.

Flowers appear from April to July. The spiny pods with long hooked bristles are unmistakable. Dispersal is by water and furry animal. Mats in animal fur can persist until the next shedding cycle. The mats are troublesome in the fine fur of house cats and some dogs. The burs do not appear to be as persistent in the coarser hair of javelina and deer (Fig. 9.78).

Moore reports several medicinal uses for Cocklebur and says tea made by boiling two or three pods will stop the most obstinate diarrhea. However, too frequent use can be toxic for the intestinal tract and liver (Moore 2003).

Fig. 9.78 Cocklebur growing on the floodplain of the Agua Fria River (photo © GR)

Fig. 9.79 Common Sagewort (*Artemisia campestris*)—ASTERACEAE. Aromatic herbs with 1–5 erect stems that grow to 5 ft. tall. **Stems and Leaves:** The stems are ribbed and brown with alternate pinnately divided leaves. **Flowers**: Tiny yellow flowers occur in spikelike panicles (Fig. 9.79 painting © Kurt Stüber, Public Domain)

Common Sagewort (*Artemisia campestris*) [BP|EMN]

Sagewort is a slender aromatic herb colonizing disturbed areas throughout North America, Europe, Africa, and Australasia. Several subspecies and variants have evolved across that wide range. The native species growing in CWF has a persistent basal rosette and is the only perennial subspecies of the group (Fig. 9.80).

People have long used the Artemisias for food, perfume, and medicine. SEINet describes some of the uses, and Moore covers many uses for other species of the genus.

Fig. 9.80 Common Sagewort (photo © Matt Lavin)

Fig. 9.81 Musk Thistle (*Carduus nutans*)—ASTERACEAE. Annual solitary stems 30–200 cm tall form dense stands. Stem with spiny wings from leaf bases. **Leaves**: Alternate, linear to elliptical, 10–40 cm long, and 5–200 mm wide. Upper leaves smaller, smooth above, hairy beneath. Margins pinnately lobed and spiny toothed. **Flowers**: Heads 7–12 mm long, disk flowers 20–40 appear July–October. **Fruits**: Achene 2–5 mm long light brown, pappus bristles 11–13 mm long and minutely *barbed*. Similar *Cirsium* species have pappus of softer unbarbed bristles (Fig. 9.81—Amédée Masclef—Atlas des plantes de France. 1891. Public Domain)

Musk Thistle (*Carduus nutans*) [AB|FIT]

A European native, Musk Thistle is a serious threat to North American native vegetation. It colonizes disturbed sites along roads and across grazed fields. It spreads into surrounding vegetation and forms dense impenetrable stands in a variety of habitats and vegetation types. Botanists suspect multiple introductions of this thistle have produced multiple hybrids. This makes identification a little difficult. I have not found the weed in CWF, but it is a likely future invader (Fig. 9.82).

Fig. 9.82 Musk Thistle (photo © Kristian Peters)

Fig. 9.83 Oxeye Daisy (*Leucanthemum vulgare*)—ASTERACEAE. Perennial 10–100 cm tall reproducing from seeds and from rhizomes; flowers spring to fall. **Leaves**: Basal leaves 12–50 mm long, 8–30 mm wide, 10–120 mm stalks, toothed with 3–7 shallow lobes. Upper leaves 30–80 mm long and 2–15 mm wide some toothed and some without stalks. **Flowers**: Solitary at ends of branches, yellow disk 1–2 cm wide, white rays 15–35 mm long, and 1–2 cm wide. **Fruit**: Achenes with 10 ribs (Fig. 9.83 photo © Max Licher)

Oxeye Daisy (*Leucanthemum vulgare*) [P|FI]

People often plant this Eurasian weed for its showy flowers. It is also an excellent source of nectar and pollen for bees, butterflies, wasps, flies, and other arthropods. In America, however, it can spread from gardens into native grasslands and form dense colonies. Cattle do not eat Oxeye Daisy, but their grazing and trampling create sites for it to colonize. Once established, it spreads by underground rhizomes. I do not know how the rhizomes respond to fire, but they can survive in soil scooped up in earthmoving operations. If they can survive fire, the plant could become one of our most troublesome weeds (Fig. 9.84).

Fig. 9.84 Oxeye Daisy © Derek Ramsey

Fig. 9.85 Welted Thistle (*Carduus acanthoides*)—ASTERACEAE. Biennial 30–200 cm tall. Spiny wings extend down along stem from leaf bases. **Leaves**: Spiny toothed, alternate 10–40 cm long, 1–20 cm wide. Upper blades smaller, hairy underneath. **Flowers**: Several heads in panicle-like array, involucre 7–12 mm long, disk 4–8 cm wide. Blooms July–October. Distinguished from similar species by having *minutely barbed bristles surrounding the flower*. **Fruits**: Achene 2–5 mm long light brown, pappus bristles 11–13 mm long. (Fig. 9.85 Photo © Jacob Sturm (1796))

Welted Thistle (*Carduus acanthoides*) [B|FI]

This Eurasian weed is widespread in the Western United States. Its plumed seeds dispersed by wind colonize disturbed sites across many types of habitats. I have not identified it yet in CWF, but it is probably here (Fig. 9.86).

Fig. 9.86 Welted Thistle (photo © ANRo0002)

Fig. 9.87 Scotch Thistle (*Onopordum acanthium*)—ASTERACEAE. Tall biennial (50–400 cm), with tomentose spiny-winged stems and leaves. **Leaves**: Alternate, 10–60 cm long, 2–15 cm wide. Basal leaves with stems, upper leaves without stems but with wings that flow down along stems. **Flowers**: Heads solitary or numerous with 18–50 blue ray flowers. **Fruits**: Achene, 4–5 mm long brown or black, pappus with many pink to red bristles. Plants produce 8000–40,000 seeds. In Scotch Thistle, the receptacle or pedicel, the thickened part of the stem at the base of the flower, is honeycombed (Fig. 9.87 painting: Otto Wilhelm Thomé, *Flora von Deutschland,* Gera)

Scotch Thistle (*Onopordum acanthium*) [B|ITZ-2]

Here is a challenge: My neighbor swore Goldfinches do not eat the seeds of this thistle, but I have not been able to observe the plants long enough to decide this is true. However, if you have none, do not plant any; Scotch Thistle is so invasive and hard to eradicate, the State of Arizona has declared it a restricted noxious weed (Fig. 9.88).

Fig. 9.88 Scotch Thistle (photo © Benutzer: Drahkrub)

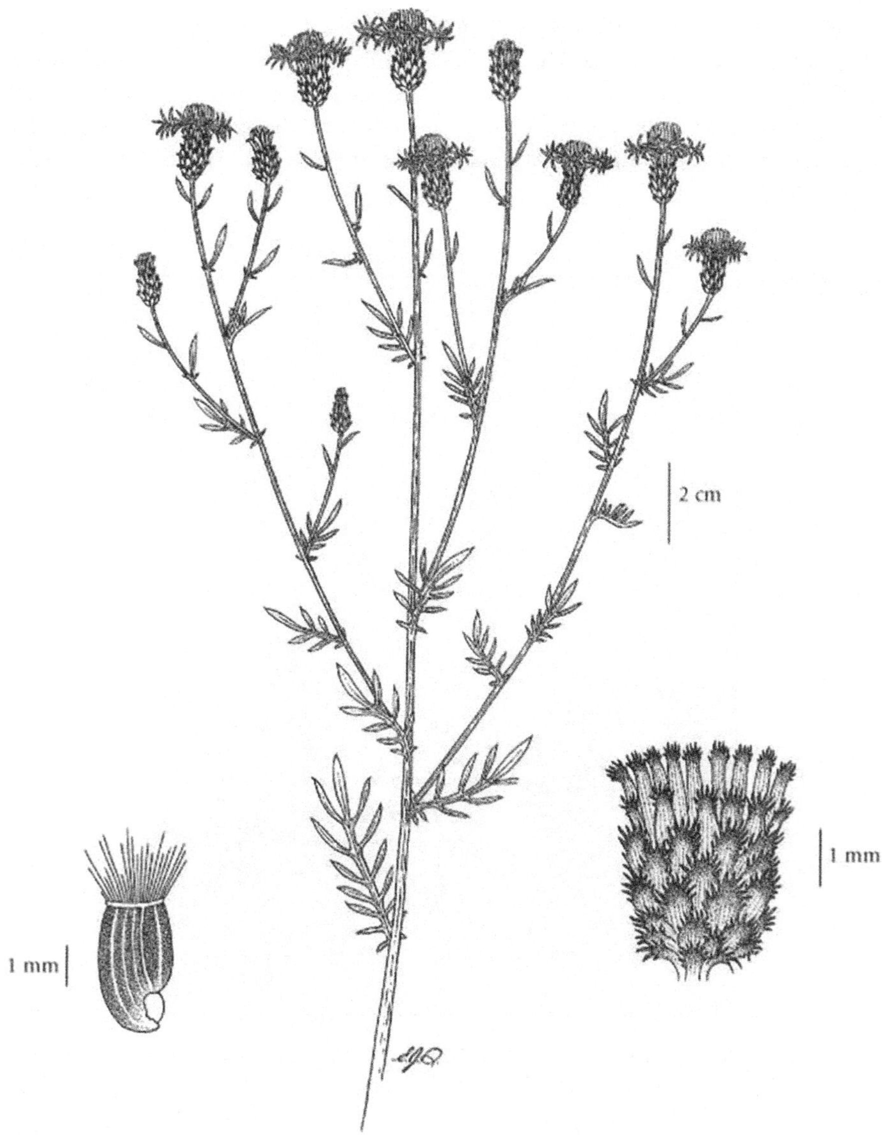

Fig. 9.89 Spotted Knapweed (*Centaurea biebersteinii*)—ASTERACEAE. Biennial 30–150 cm tall. **Leaves**: Pale gray-green, pinnate, alternate, up to 6 in. long. **Flowers**: Purple-pink at branch tips, involucre 4–13 mm. **Seeds**: Achene about 3 mm long with bristly pappus. Inset left, achene with short pappus. Inset right, involucre showing dark tips on middle and upper bracts. (Artist unknown. Internet image with no active links (Fig. 9.89). Searches across botanical artists produced the possibility the artist was Thomas Jenkinson Woodward)

Spotted Knapweed (*Centaurea biebersteinii*) [P|ITZ-2]

This eastern European weed merits the most stringent vigilance and eradication by hoeing and pulling. Its wind-blown seeds germinate in disturbed areas from whence it spreads into yards and gardens and native vegetation. It produces and disperses thousands of seeds, it develops a taproot that absorbs moisture faster than its neighbors, it produces toxic chemicals to inhibit neighboring plant germination and growth, and it develops foliage of low appeal to cattle. Sheep can eradicate this weed, but if you do not have sheep, get the hoe (Fig. 9.90).

Fig. 9.90 Spotted Knapweed (photo © Matt Lavin)

Fig. 9.91 Yellow Starthistle (*Centaurea solstitialis*)—ASTERACEAE. Annual, bushy, gray hairy plants to 1 m tall. **Leaves**: Basal lobed 20–50 cm, become smaller and unlobed with height on plant. **Flowers**: Yellow; involucre 10–15 mm wide at base; middle and outer bracts with spines 11–22 mm long. **Fruits**: Dark brown achenes 2–3 mm long (Fig. 9.91 photo Peggy Greb, ARS, USDA)

Yellow Starthistle (*Centaurea solstitialis*) [A|IPZ-2]

With its impressive thorns and beautiful flowers, this foreign weed might tempt you to give it a spot in your yard. Not a good idea. Though the flowers attract butterflies, the plant is toxic to soil microorganisms and other plants, and both its roots and seeds are persistent. Mowing and burning often do not control the plant's spread. Herbivorous insects and competing plants keep it in check in its homeland in Eurasia, but in New World deserts, the controls are not present; it invades and replaces native vegetation (Fig. 9.92).

Fig. 9.92 Yellow Starthistle (photo © Franco Folini from San Francisco)

Fig. 9.93 Stinking Chamomile (*Anthemis cotula*)—ASTERACEAE. Annual 5–90 cm tall, green stems dotted with stinky glands. **Leaves**: Pinnately lobed 25–55 mm long, 15–30 mm wide. **Flowers**: Heads solitary; male and female flowers on separate plants; disk flowers only, yellow centers, white petals. **Fruit**: Achene 1–2 mm long (Fig. 9.93 painting, Norman Criddle)

Stinking Chamomile (*Anthemis cotula*) [A|IM]

Also known as Dog's Fennel, this is not the plant used to make tea. Lots of insects visit the flowers, and some of the sources list medicinal properties. However, Moore does not cover it, so you will have to research the topic if interested in medical applications (Fig. 9.94).

Fig. 9.94 Stinking Chamomile (photo © Max Licher)

Fig. 9.95 Sweet Resinbush (*Euryops multifidus*)—ASTERACEAE. Bright green shrub to 1 m tall. **Leaves**: To 2 cm long, crowded near stem tips; some divided into narrow lobes about 1 mm wide, fleshy and smooth. **Flowers**: About 1.5 cm wide, yellow ray and disk flowers on leafless stalks extend to 15 cm from stem tips. **Seeds**: Hairy and when wet secrete a gelatin-like substance sticks to feathers, fur, machines, and clothing (Fig. 9.95 photo, U.S. Forest Service)

Sweet Resinbush (*Euryops multifidus*) [P|FITZ-2]

Sweet Resinbush (Fig. 9.96) poses a serious threat to native vegetation in North American deserts. It is a small shrub with a sticky substance on the branches. Hoping to have found a better cattle food than native plants, range scientists introduced it from South Africa in the 1930s. When cows refused to eat it, they abandoned the project. More than 30 years later, Sweet Resinbush began spreading.

"Why the concern over this shrub? Contrary to the intended goals of increasing forage for livestock and reducing erosion, sweet resinbush does the opposite. The plant encroaches into healthy grasslands, choking out native vegetation. It then forms a monoculture of resinbush with large amounts of bare soil, leading to increased soil erosion" (McReynolds 2008, p. 9).

This plant is now growing near Prescott northwest of CWF and in the Agua Fria National Monument south of CWF. It could reach CWF soon.

Fig. 9.96 Sweet Resinbush (*Euryops multifidus*) (photo © E. Makings)

9.5 Forget-Me-Not Family: BORAGINACEAE (Fig. 9.97)

Fig. 9.97 Coast Fiddleneck (*Amsinckia intermedia*)—BORAGINACEAE. Stiff harsh hairs cover the erect, branching stems 6–30 in. high and the 1–6 in. long leaves. **Flowers**: Slender yellow or orange-yellow, 9–13 mm long, crowded on one side of the 3–10 inch long curling spikes. The bristly yellow five-lobed calyx encloses four gray nutlets containing one seed. **a**. Nutlet (Fig. 9.97 from *An Illustrated Guide to Desert Weeds*, by Kittie F. Parker. © 1972 The Arizona Board of Regents. Reprinted by permission of the University of Arizona Press)

Coast Fiddleneck (*Amsinckia intermedia*) [A|N]

This native annual weed appears in all the usual places: roadsides, gardens, and other disturbed areas. Mowing, hoeing, or pulling before seeds develop will soon eradicate it. Despite all the hairs, animals eat Coast Fiddleneck and Gila River tribes ate the leaves after boiling, straining, and refrying them. The seeds are poisonous (Fig. 9.98).

Fig. 9.98 Coast Fiddleneck (photo © Curtis Clark)

Fig. 9.99 Alkali Heliotrope (*Heliotropium curassavicum* var. *oculatum*)—BORAGINACEAE. **a**. Pair of flower spikes. **b**. Young fruits. **c**. Fruit composed of four small brown nutlets 1–2 mm long. **d**. Single nutlet. Older flowers are at the fork of the uncoiling spikes (Fig. 9.99 from *An Illustrated Guide to Desert Weeds*, by Kittie F. Parker. © 1972 The Arizona Board of Regents. Reprinted by permission of the University of Arizona Press)

Alkali Heliotrope (*Heliotropium curassavicum* var. *oculatum*) [P|DFN]

A native perennial, Alkali Heliotrope (AH) has tiny fragrant flowers that draw bees and gnats. The uncoiling flower-studded fiddleneck reveals tiny flowers with yellow-green eyes and pure white petals that darken to purple in sunlight. This low perennial grows to about one foot high in patches of overlapping branches up to 4 ft. wide. The plants are hairless and dusted with white powder.

A native of arid North America, AH can edge into a lawn or garden. Removal by pulling is a simple cure. The plant has no thorns or burrs and you might wish to leave it on unused sites where it will protect your soil from wind and water erosion.

The plants colonize exposed alkaline or saline soils and the banks of streams and washes in arid Western United States and Baja California. You will find similar varieties around the world. I have seen this one growing beside a canal in Tempe, Arizona and similar one beside limestone rocks near Fourteen Mile Creek in Oklahoma (Fig. 9.100).

Fig. 9.100 Alkali Heliotrope flowers (photo © David Eickhoff)

9.6 Mustard Family: BRASSICACEA (Fig. 9.101)

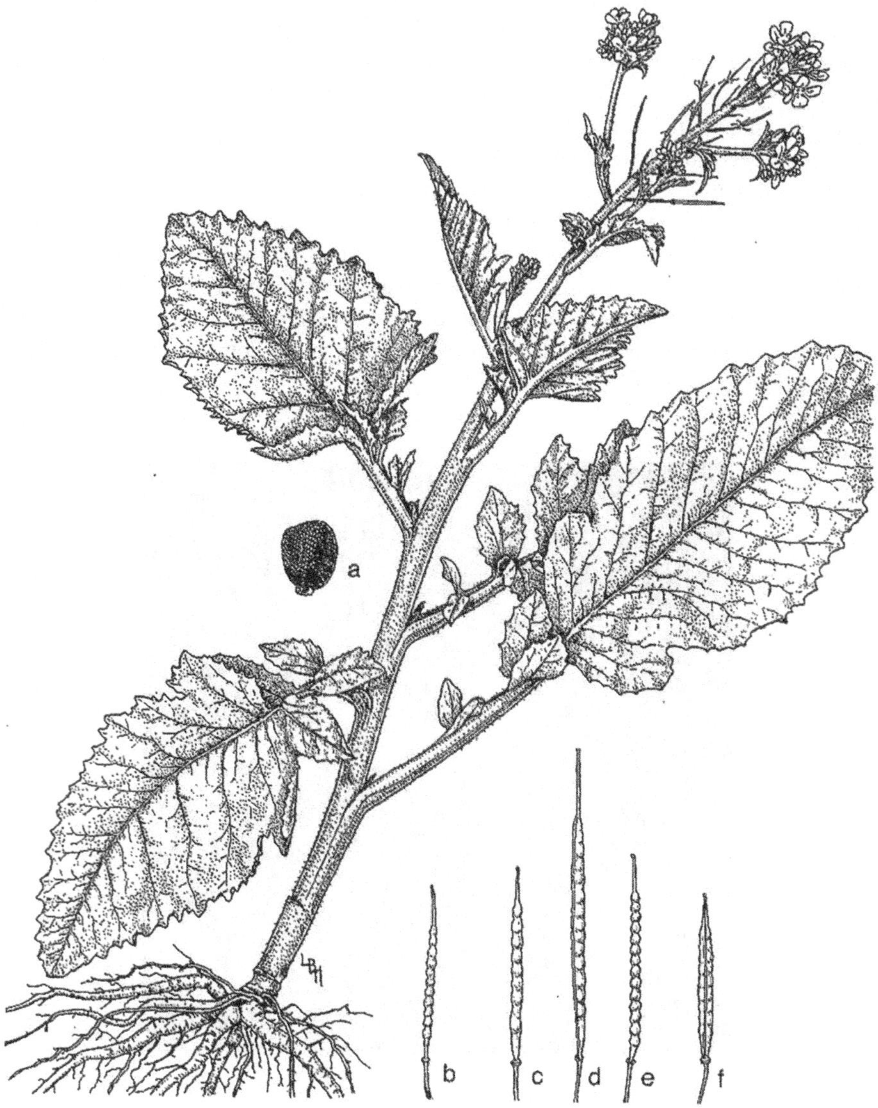

Fig. 9.101 Black Mustard (*Brassica nigra*)—BRASSICACEA. Annual, 2–6 ft. tall, reproduces by seed. Lower leaves are 4–16 inches long and have pairs of small lobes. Higher leaves are small, unlobed, and curled downward. Bright yellow flowers with four petals on the upper 1–2 ft. of the stem. The siliques (seedpods) are 9–19 mm long at maturity and about 2 mm wide on short stalks. Seeds are 1–2 mm long and brown to black. **a**. Seed. **b**. Silique. **c**. Indian Mustard (*B. juncea*) silique. **d**. Wild Turnip (*B. campestris*) silique. **e**. Wild mustard (charlock) (*B. kaber*) silique. **f**. Rocket salad (*B. eruca*) silique (Fig. 9.101 from *An Illustrated Guide to Desert Weeds*, by Kittie F. Parker. © 1972 The Arizona Board of Regents. Reprinted by permission of the University of Arizona Press)

Black Mustard (*Brassica nigra*) [A|EI]

Black mustard originated in the Middle East, but farmers grow it worldwide to produce yellow table mustard. Its relative, Indian Mustard (*B. juncea*) is the source of brown or Dijon mustard. The leaves add tang to salads and steamed vegetables. Both mustard species will spread into other crops and along streams and ditches beside roads. Black Mustard is the most likely to appear in CWF (Fig. 9.102).

Checkered White (*Pontia protodice*), butterflies lay their eggs on this weed.

Fig. 9.102 Black Mustard ripening siliques and flowers (photos © Pancrat)

Fig. 9.103 Blue Mustard (*Chorispora tenella*)—BRASSICACEA: Annual, 2–24 inches tall, around seven stems upright or decumbent and branched. **Leaves**: Alternate, oblong, 1–13 cm long. **a**. Groovy stem. **b**. **Flower**, lavender to bright purple (rarely white), petals narrow, 8–20 mm long, sepals forming a deep purple tube. **c**. Six stamens with two shorter, style same length as filaments. **d**. **Silique** cylindrical, slightly curved, spreading, 12–25 mm long (Fig. 9.103, Flora Batava 20, 1898)

Blue Mustard (*Chorispora tenella*) [P|IP]

This Asian annual is one of the most common weeds around CWF. It appears in early spring and soon fills the air with its "perfume" which some people find repugnant. The plant forms large patches along our streets and sometimes it covers yards and vacant lots. Standing downwind of such a stand can produce a hint of nausea. Though the plant is small and produces no thorns or burrs, I consider it one of my worst weeds and give it the symbol **P**. Every year, I find and pull the scouts around the house. That helps, but when the wind is right, the plants up the street fill my air with a fragrance not unlike burning plastic (Fig. 9.104).

Fig. 9.104 Blue Mustard (photo © Stan Shebs)

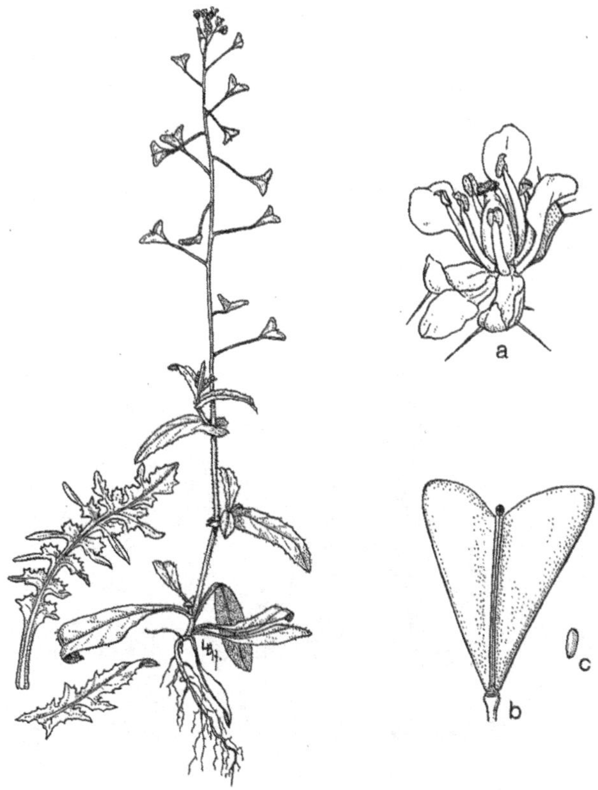

Fig. 9.105 Shepherdspurse (*Capsella bursa-pastoris*)—BRASSICACEA. Annual 3–18 in. high, reproducing only by seed. Slender stems with long gray hairs. **Leaves**: Spreading rosette first to form. Rosette leaves toothed or lobed 4–8 cm long including stalk. Stem leaves alternate, not lobed, pointed, stalkless, clasp stem with earlike lobe on both sides. **Flowers**. White, 2–3 mm long on slender stalks. **Seeds**: In triangular silicles (seedpods with width less than three times length) notched with pointed base, about 6 mm wide. Reddish or orange brown seeds are oblong, shiny, about 1 mm long, grooved on each side. **a**. Flower. **b**. Seedpod. **c**. Seed (Fig. 9.105 from *An Illustrated Guide to Desert Weeds*, by Kittie F. Parker. © 1972 The Arizona Board of Regents. Reprinted by permission of the University of Arizona Press)

Shepherdspurse (*Capsella bursa-pastoris*) [A|M]

The distinctive triangular silicles of Shepherdspurse make identification as easy as it gets. The plant has a long flowering and fruiting season from spring to fall and some silicles will persist into winter (Fig. 9.106).

All mustard seeds are edible, but these are the best. However, the plant has strong chemical constituents making it unwise to eat more than a few of the fresh silicles. This caution does not limit Checkered White, Cabbage White, and Sara Orange-tip butterflies using the plant to host their caterpillars.

Fig. 9.106 Shepherdspurse painting by Norman Criddle

Fig. 9.107 Hoary Cress (*Lepidium draba*)—BRASSICACEA: Gray-green perennial 6–24 in. high reproduces by seed and sprouts from deep, spreading roots. Erect stems branch to form floral bouquets. **Leaves**: Broad, elliptical, 1–4 in. long. Lower leaves stalked, upper leaves stalkless, clasping stem with two lobes. **Flowers**: Tiny flowers on spreading stalks 6–13 mm long have 3 mm **long pet**als. **Seeds**: Seedpods are almost hairless 3–6 mm wide, tipped by slender beak. Two chambers produce one or two seeds 1–2 mm wide. **a**. Lower leaf. **b**. Seedpod. Hairy White-top (next) is very similar (Fig. 9.107 from *An Illustrated Guide to Desert Weeds*, by Kittie F. Parker. © 1972 The Arizona Board of Regents. Reprinted by permission of the University of Arizona Press)

Hoary Cress (*Lepidium draba*) [P|I]

This medium-sized alien perennial prefers the dry air of the desert. First collected in the USA on Long Island in 1862 (Groh 1940), Hoary Cress is abundant in the Rocky Mountain States and westward. The plant is rare in CWF.

People introduced Hoary Cress on Long Island and elsewhere as an ornamental. They also introduced it by accident when they imported contaminated Alfalfa seeds. The plant is most successful in the sagebrush vegetation of the northern Great Basin Desert. Part of its success is because of toxic chemicals in the roots. These saturate the soil around the plant and inhibit germination and seedling growth of other plant species. Hoary Cress can replace native vegetation (Fig. 9.108).

Fig. 9.108 Hoary Cress (photo © Kraelj)

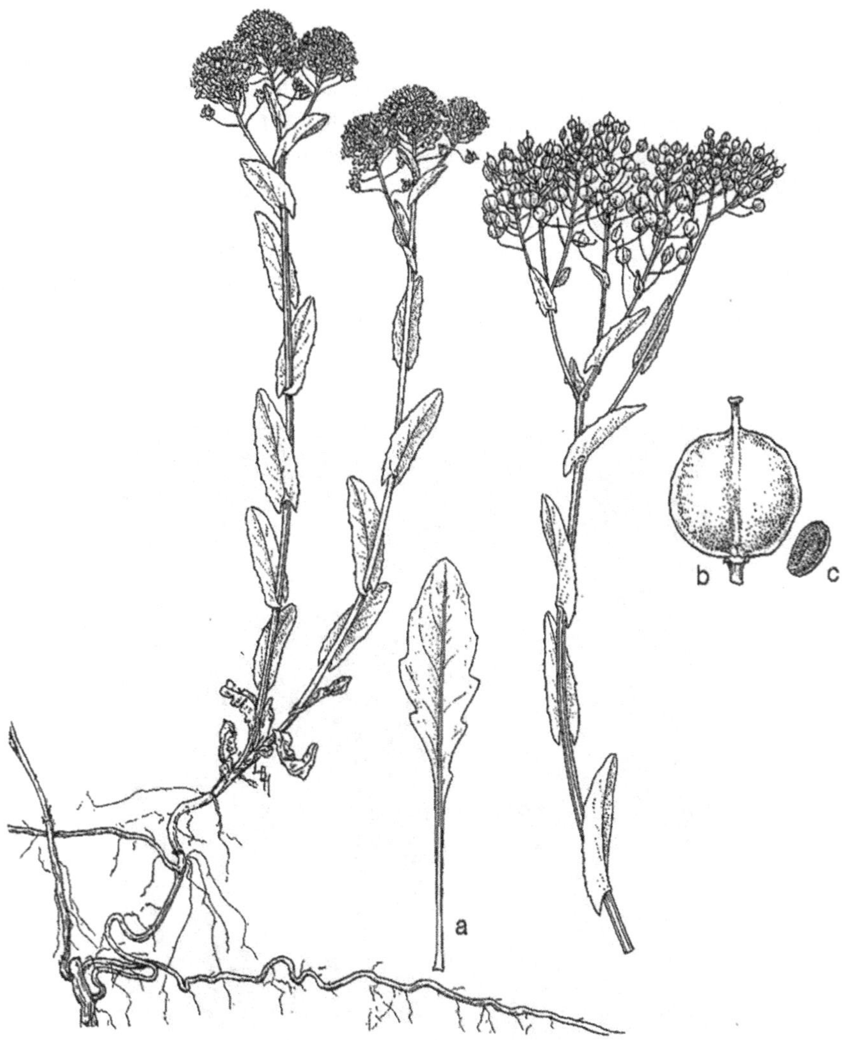

Fig. 9.109 Hairy White-top (*Lepidium appelianum*)—BRASSICACEA: Gray-green perennial 6–24 in. high reproduces by seed and sprouts from deep, spreading roots. Erect stems branch to form floral bouquets. **Leaves**: Broad, elliptical, 1–4 in. long. Lower leaves stalked, upper leaves stalkless, clasping stem with two lobes. **Flowers**: Tiny flowers on spreading stalks 6–13 mm long have 3 mm long petals. **Seeds**: Seedpods are almost hairless 3–6 mm wide, about 8 mm long and tipped by slender beak. Two chambers produce one or two seeds 1–2 mm in size. **a**. Lower leaf. **b**. Seedpod shown is from the nearly identical Hoary Cress (*Cardaria draba*) (cf.). *Hairy White-top seedpods are hairier than those of Hoary Cress* (Fig. 9.109 from *An Illustrated Guide to Desert Weeds*, by Kittie F. Parker. © 1972 The Arizona Board of Regents. Reprinted by permission of the University of Arizona Press)

Hairy White-top (*Lepidium appelianum*) [P|FI]

This Asian introduction was unknown in Arizona when Kitty Parker published her *Arizona Weeds* (Parker 1972). According to the Encyclopedia of Life, this Uzbekistan native was present on Long Island, New York in 1871. The plant did not spread to other eastern or Midwestern states. However, it is now present in the Rocky Mountains, Great Basin, and arid eastern portions of the West Coast states.

The plant is present in one location north of the Grand Canyon and in Skull Valley west of Prescott, Arizona. Botanists surveying the Agua Fria National Monument in 2007 found the plant and judged it a high threat for invasion. The Monument is about 20 miles south of CWF, close enough to expect any weeds there will soon be here.

The plant's absence in the Great Plains suggests it requires low humidity to thrive. We have that (Fig. 9.110).

Fig. 9.110 Hairy White-top (photo © Meneerke Bloem)

Fig. 9.111 Tansymustard (*Descurainia pinnata*)—MUSTARD FAMILY—BRASSICACEA. Annual reproducing only by seed. Pale gray due to covering of forked or starlike hairs. Some less hairy and green. Stems often purple. **Leaves**: Alternate, 1–4 in. long, divided into fine segments. **Flowers**: Pale yellow, 2 mm long on short stalks which lengthen as siliques mature. **Seeds**: Short seedpods, 6–13 mm long, on upper stems; contain two rows of dull red seeds about 1 mm long with a grove on one side. The siliques are wider and rounded at the tip than the base (club-shaped), *siliques not constricted between the seeds*, pods point upwards. **a**. Portion of seedpod showing seed attachment. **b**. Seedpod. **c**. Seed (Fig. 9.111 from *An Illustrated Guide to Desert Weeds*, by Kittie F. Parker. © 1972 The Arizona Board of Regents. Reprinted by permission of the University of Arizona Press)

Tansymustard (*Descurainia pinnata*) [A|EIN]

Though a native weed, Tansymustard (Fig. 9.112) poses great danger to native ecosystems. It combines with other weeds to increase fine fuel quantities and encourage increased wildfire frequency. If the interval between fires becomes too short, the native plants normally succeeding Tansymustard and other colonists cannot return. Perhaps this is only when other fuelweeds are present, perhaps not. I know of one instance where Tansymustard was the principal weed replacing an extensive Greasewood (*Sarcobatus vermiculatus*) community. The plant is abundant mixed with other weeds on Black Mesa in Agua Fria National Monument south of CWF. Over the past 50 years, Black Mesa fire frequency increased eliminating all native vegetation.

Fig. 9.112 Tansymustard (photo © Curtis Clark)

Fig. 9.113 Flixweed (*Descurainia sophia*)—BRASSICACEA: Introduced annual 25–75 cm tall. **Leaves**: Basal leaves with short stalks; stem leaves with no, or with short, stalks, alternate, 2–9 cm long, 2–3 times pinnate. **Flowers**: Yellow, terminal racemes on 8–15 mm long pedicels ascending in flower and spreading to descending when in fruit; four sepals ascending, yellowish, 2–4 mm long; petals 2–3 mm long. **Seeds**: Many brown 1 mm long seeds in 1–3 cm long, curved, ascending siliques *constricted between the seeds*. **3**. Basal stem and pediceled leaf. **4**. Stem. **5**. Flower showing curving sepals. **6**. Silique. **7**. Seed. (Fig. 9.113 painting: Lindman 1917–1926)

Flixweed (*Descurainia sophia*) [A|I]

Flixweed is common across the Western United States. My original encounter with this weed occurred while researching historical vegetation changes in the Great Basin Desert. There, the weed was quick to arrive after a fire. Like Tansymustard (above), the plants dried out by early summer served as fuel to carry fires. Fire frequency increased, and in many places, none of the original vegetation remains (Fig. 9.114).

Fig. 9.114 Flixweed forming a fine fuel matrix between shrubs 2 years after a fire (photo © Max Licher)

Fig. 9.115 Perennial Pepperweed (*Lepidium latifolium*)—BRASSICACEA. Rhizomatous perennial to 2 m tall. **Leaves**: Lower leaves leathery on stalks 1–10 cm, oblong, 10–30 cm long, and 2–8 cm wide. Upper leaves smaller. **Flowers**: Petals white 2–3 mm long, 1 mm wide, six stamens. Blooms June–August. **Seeds**: Oblong, about 1 mm in size (Fig. 9.115 Flora Batava 20, 1898)

Perennial Pepperweed (*Lepidium latifolium*) [P|EFI]

This is another Eurasian weed spreading across North American Deserts. I have not yet seen it in CWF. Its most likely first appearance will be along the Agua Fria River since it prefers moist soil (Fig. 9.116).

Perennial Pepperweed can invade desert grasslands and shrublands. Since the shoots, leaves, and seeds are nutritious (boil, soak for two days, then treat like spinach) we could arrest this invader's spread on the home cookstove.

Fig. 9.116 Perennial Pepperweed (photo © Michael Becker)

Fig. 9.117 Hairy-Pod Pepperwort (*Lepidium lasiocarpum*)—BRASSICACEA. Annual 1–15 in. tall, stems erect or prostrate, numerous, with short hair lying flat on surfaces, sometimes with longer hair. **Leaves**: 1–6 in. long, alternate, wider toward tip, toothed or lobed. **Flowers**: White petals less than 2 mm long or absent supported by hairy pedicel. **Seeds**: Rounded seedpods notched at tip covered by short hair or by hair on edges, 3–16 mm long by 3 mm wide and contain two red, oval seeds. **a**. Flower. **b**. Seedpod with short hairs. **c**. Seed. Subsp. *wrightii* seedpods have pustular-based hairs, and subsp. *lasiocarpum* seedpod hairs are not pustular based. (Fig. 9.117 from *An Illustrated Guide to Desert Weeds*, by Kittie F. Parker. © 1972 The Arizona Board of Regents. Reprinted by permission of the University of Arizona Press)

Hairy-Pod Pepperwort (*Lepidium lasiocarpum*) [A|EN]

With your hand lens, you can see the *hairy flattened pedicels and seed stalks* that distinguish this annual native from other mustards. The species name, *lasiocarpum* means woolly seeds or fruits. In fact, some botanists call the plant Shaggyfruit Pepperweed (Fig. 9.118). You might form your own preferred name by combining parts of these. According to Gledhill (1985), *lasio* is the Latin prefix meaning woolly or shaggy, so I guess you could substitute woolly without creating too much confusion. *Carpum* of course means fruit (Fig. 9.118).

Fig. 9.118 Hairy-Pod Pepperwort, subsp. *lasiocarpum*. (photo © Stan Shebs)

Fig. 9.119 Dyer's Woad (*Isatis tinctoria*)—BRASSICACEAE. Biennial to 14–36 in. high. **Leaves**: Basal leaves hairy, 2–7 in. long, 1–4 cm wide; stem leaves 1–4 in. long. **Flowers**: Yellow, 6 mm wide with 3 mm long petals. Born in compound racemes in a large terminal panicle. **a**. Lower portion of plant. **b**. Upper portion with flowers. **c**. Seeds. **1**. Flower. **2**. Flower cross section. **3**. Six stamens surrounding stigma. **4**. Achene. **5**. Achene bisected. **6**. Seed (Fig. 9.119 Amédée Masclef—Atlas des plantes de France. 1891)

Dyer's Woad (*Isatis tinctoria*) [B|FI]

Not yet seen in CWF, Dyer's Woad (Fig. 9.120) is spreading across the region and could reach the farm soon. People cultivated the weed for thousands of years for indigo dye. In the early 1900s, synthetic dyes replaced Woad. It is an attractive plant, but there are reports it is spreading into grazed rangelands and might become a major problem (Whitson et al. 2006).

Fig. 9.120 Dyer's Woad (photo © Stefan Lefnaer)

9.7 Hackberry Family: CANNABACEAE (Fig. 9.121)

Fig. 9.121 Netleaf Hackberry (*Celtis reticulata*)—CANNABACEAE. Tree with round crown 1–16 m tall, trunk less than 1 m thick. **Leaves**: Smooth margins, pointed 1–4 cm wide, 2–5 cm long with prominent veins. Cystoliths (bumps on leaves formed of cells with calcium carbonate accumulations and surrounded by epidermal cells) are common. **Flowers**: Appear with leaves in early spring, 1–4 in leaf axils. **Fruit**: Drupes red, blackening as they ripen, oblate spheroidal, 5–10 mm in diameter (Fig. 9.121 photo © Max Licher)

Netleaf Hackberry (*Celtis reticulata*) [P|DEN]

This native tree of the North American deserts and arid western Great Plains has remained at home. Deer and several insect and moth species eat Hackberry leaves, and birds eat the fruit. Mature trees are usually less than 30 ft. tall. They are drought tolerant and can suppress sub-canopy weeds. They make fine additions to CWF flora (Fig. 9.122).

Fig. 9.122 Netleaf Hackberry with ripening berries (photo © GR)

9.8 Honeysuckle Family: CAPRIFOLIACEAE (Fig. 9.123)

Fig. 9.123 Common Teasel (*Dipsacus fullonum*)—CAPRIFOLIACEAE. Three to eight feet tall, prickly all over. **Leaves**: Lanceolate, 8–16 in. long, and 3–6 cm wide with a row of small spines on the underside of the midrib. **Flowers**: Inflorescence: ovoid to cylindrical, 1–4 in. long. Involucral bracts are unequal, curving upwards, often surpassing the heads. Calyx silky, 1 mm long, corolla 8–19 mm, white with purple lobes 1 mm long. **Seeds:** 5 mm long mature in autumn retaining the calyx (Fig. 9.123 painting Johann Georg Strum (1796))

Common Teasel (*Dipsacus fullonum*) [B|FI]

Teasel is a European biennial. It is not common in our area, but its distinctive inflorescence will catch your attention. At higher elevations, this plant can invade native vegetation and crowd out all native plant species.

Navajos use the dry inflorescence to card wool, and people often add it to dried flower bouquets. Teasel is weakly carnivorous. The leaf axels form cups that hold water. When scientists added insects to the cups, the plants produced more seeds (Fig. 9.124).

Fig. 9.124 Flower (photo © AnemoneProjectors)

9.9 Morning Glory Family: CONVOLVULACEAE (Fig. 9.125)

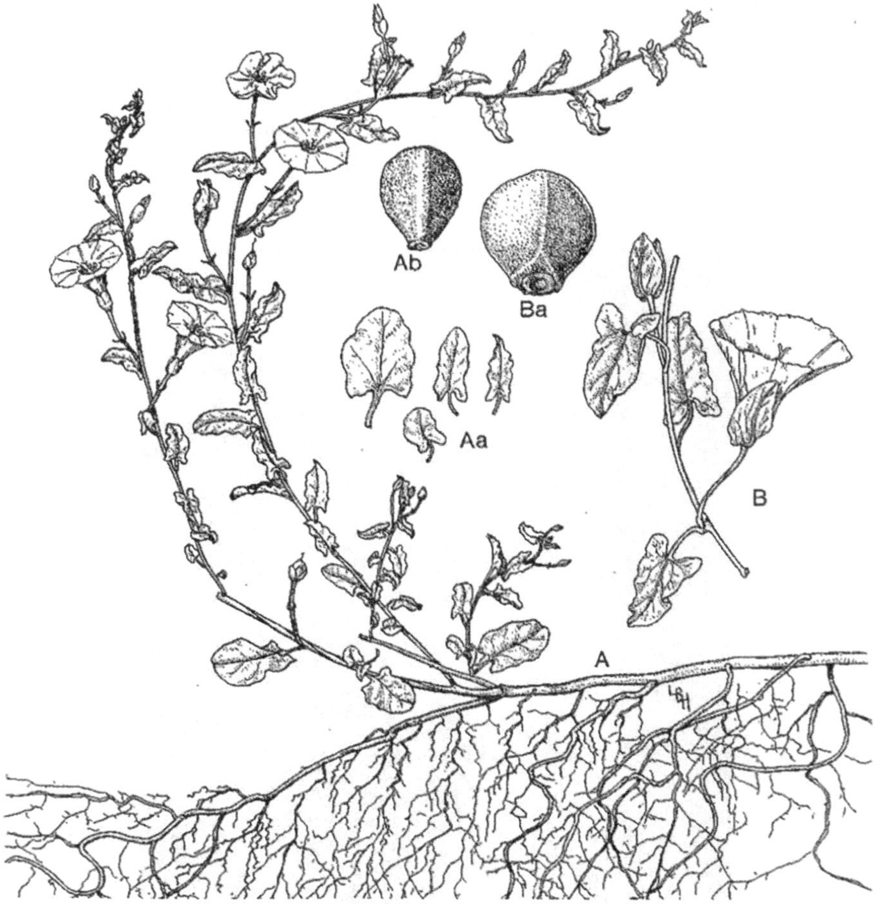

Fig. 9.125 Bindweed (*Convolvulus* spp.)—CONVOLVULACEAE. Field Bindweed's (*Convolvulus arvensis*) slender stems spread 1–10 ft. from a deep taproot. They climb fences and other plants. **Leaves**: Eggshaped, lobed at base, 2–5 cm long alternate on 6–20 mm long stalks. **Flowers**: Funnelshaped, white or pink, 2–3 cm long, and 2–3 cm wide arising from leaf axils. A pair of narrow pointed bracts, 3–5 mm long occur on the stalks. There are five sepals, *five fused petals, and five stamens that fuse with the petals*. **Seeds**: Seedpods are 4–6 mm long and contain 1–4 dark brown or black eggshaped seeds 3–5 mm long. **a**. Field Bindweed. **Aa**. Various shapes of leaves. **Ab**. Seed. Hedge Bindweed (*Convolvulus sepium*) resembles Field Bindweed, but **leaves** are longer, 6–9 cm, and *the two leaf stalk bracts are much larger*, 13–25 mm long, and just below the flower. The **flowers** are larger, 4–6 cm long. **b** Hedge Bindweed branch with flower. Ba. Seed.Black Bindweed (*Polygonum convolvulus*) leaves resemble Field Bindweed but the flowers are tiny and green or tinged purple (Fig. 9.125 from *An Illustrated Guide to Desert Weeds*, by Kittie F. Parker. © 1972 The Arizona Board of Regents. Reprinted by permission of the University of Arizona Press).

Bindweed (*Convolvulus* spp.) [P|TZ-1]

One Master Gardener I know says Bindweed is her greatest problem. Wherever and whenever disturbances expose soil on roadsides and in cultivated fields, yards, gardens, and even native woodlands, Bindweed will be there. The plant does best where there is something to climb including fences, stakes, and other plants. On other plants, it can get so dense it hinders growth. The plant is very difficult to eradicate; it forms deep root systems from which it sprouts anew every spring (Fig. 9.126).

Fig. 9.126 Bindweed climbing over itself struggling to reach the glorious heights of the 2 ft. post beside my front walk (photo © GR)

Fig. 9.127 Scarlet morning glory (*Ipomoea coccinea*)—CONVOLVULACEAE. Hairless annual with ridged, often reddish stems climbing nearby plants or structures or trailing on the ground; reproduces only by seed. **Leaves**: Alternate 4–7 cm long on 1–4 in. long stalks, may be unlobed on some plants and heartshaped with a long pointed tip on others. On some varieties (e.g., var. *hederifolia*), some or all leaves have 3–5 fingerlike lobes. **Flowers**: Bright red, trumpetshaped 2–3 cm long and 1–2 cm wide. Stalks 3–5 in. long arise in leaf axils and support two or more flowers. **Seeds**: Globular seedpods have 4–6 black seeds about 3–4 mm in diameter. **a**. Seedpod. **b**. Seed (Fig. 9.127 from *An Illustrated Guide to Desert Weeds*, by Kittie F. Parker. © 1972 The Arizona Board of Regents. Reprinted by permission of the University of Arizona Press)

Scarlet Morning Glory (*Ipomoea coccinea*) [A|DN]

This beautiful morning glory is native to the Southeastern United States. People introduced it to the Southwest where it is now present beside roads, along waterways, and in any disturbed place where there is supplemental water. It climbs the volunteer sunflowers on the edge of my back lawn, and it is present down near the river. Around the farm, you will see it in rocky depressions and beside streets. In your yard, you may find it climbing on other plants and fences (Fig. 9.128).

Fig. 9.128 Scarlet morning glory (photo © GR)

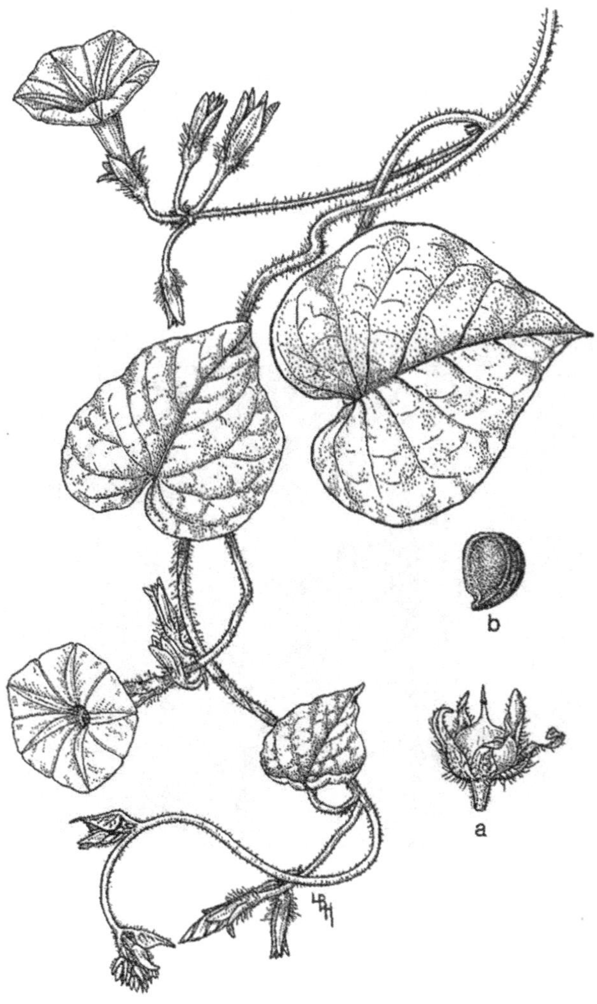

Fig. 9.129 Tall morning glory (*Ipomoea purpurea*)—CONVOLVULACEAE. Annual reproduces by seed. Climbing or trailing stems are hairy, 5–13 ft. long. **Leaves**: Heartshaped, unlobed, hairy, pointed, 2–4 in. long on stalks 2–4 in. long. **Flowers**: White to blue, purple, or bright pink with variable markings (those I have seen are pale purple); 4–8 cm long, 4–5 cm wide with a hairy five-lobed calyx 1–2 cm long. **Seeds**: Globeshaped seedpods resemble those of Scarlet morning glory and contain 4–6 flat seeds about 5 mm long. **a**. Seedpod enclosed in calyx. **b**. Seed (Fig. 9.129 from *An Illustrated Guide to Desert Weeds*, by Kittie F. Parker. © 1972 The Arizona Board of Regents. Reprinted by permission of the University of Arizona Press)

Tall Morning Glory (*Ipomoea purpurea*) [A|D]

Occupies the same habitats as Scarlet morning glory. Climbs sunflowers and other weeds to add twinkles of color amidst the green leaves. Introduced as an ornamental from Central and South America, Tall morning glory has many cultivars. My favorite is Heavenly Blue, but like most of the others, the cultivar is not as persistent as the wild original. This weed can be a problem in some crops, but not so much in yards and gardens (Fig. 9.130).

Fig. 9.130 Tall morning glory (photo © GR)

9.10 Gourd Family: CUCURBITACEAE (Fig. 9.131)

Fig. 9.131 Buffalo Gourd (*Cucurbita foetidissima*)—CUCURBITACEAE. Large vines with deep tuberous roots and coarse stems that grow 30–40 cm high and spread 10 m or more. The bulging central taproot weighs up to 150 pounds. **Leaves**: Alternate, triangular, 10–20 cm long, rough, hairy, progressively smaller toward stem tips. Each leaf paired with a thick, coiled tendril emerging from the stem just below the leaf petiole. The leaves are not lobed. **Flowers**: Large, yellow, solitary, separate male and female flowers on each plant (monoecious). Five pointed sepals much shorter than the five petals. Fruit: Spherical 6–8 cm gourds green with white stripes drying yellow. **Seeds**: Flat ovate and numerous in each gourd. Similar gourds have lobed leaves. Buffalo Gourd montage. Top left: Three colors of maturing fruit. Top right: Staminate flower. Middle: Whole plant. Bottom left: Fruit broken open to expose seeds and pulp. Bottom right: Leaves, new fruits, and whitish fruit from the previous season. (Fig. 9.131 photographs: Up, from left to right: © Paigeblue08, © Paul Asman, © Jill Lenoble, and © Clark. Middle: © Paigeblue08. Bottom left: © Charle Bernardo. Bottom right: © Paigeblue08 (montage © RoRo)).

Buffalo Gourd (Coyote Melon) (*Cucurbita foetidissima*) [P|MN]

You will often see Buffalo Gourds from your car window while speeding along one of the Southwest's highways. In winter, the ripened yellow fruit resembles spilled oranges or lemons. Closer to home, you might spot the pointed pale leaves like the ears of an underground pixy elephant reaching through surrounding plants.

An elephant indeed, Buffalo Gourd stores moisture and carbohydrates in a tuber weighing up to 150 pounds. This reserve gives plants the ability to survive drought and aboveground assaults by browsing wildlife and lawn mowers. On overgrazed cattle ranges, you will find land where nothing remains but patches of Buffalo Gourd scattered across land devoid of other plants.

The current human-enhanced drought may prove too much even for this mighty weed. The plants grow throughout Arizona and scattered about in CWF. It died back in the dryer upper pasture at CWF after I stopped irrigating. As human-exacerbated drought continues, the plant might survive in ditches beside roads and in lower flood plains where the tuberous roots can find moisture.

Controlling Buffalo Gourd is not difficult if that is something you have to do. When the first leaves appear, stop their spread by applying a sharp shovel to the rooted stem. Feeling industrious, you could excavate the tuber and permanently eliminate the plant. Or you can keep cutting the rooted stem every month through the growing season. After a few years, the plant will die.

If you are interested in wildlife gardening, train a plant along your fence or in a waste place beside your garden. But do not pick the flowers—your fingers will stink for days.

9.11 Sedge Family: CYPERACEAE (Fig. 9.132)

Fig. 9.132 Yellow Nutsedge (*Cyperus esculentus*)—CYPERACEAE. Perennial 1–3 ft. tall, triangular stem, reproducing from seed, rhizomes running above ground and small dark underground tubers or nutlets 1–2 cm long. **Leaves**: From near base, often longer than the stem, 3–8 mm wide. **Flowers**: Umbrellalike tops have thin stems radiating from tip with yellow or golden brown spikelets 6–25 mm long and 1–2 mm wide. **Seeds**: Brown 3-angled achenes about 2 mm long. **a** Flowering spikelet. **b**. Achene (Fig. 9.132 from *An Illustrated Guide to Desert Weeds*, by Kittie F. Parker. © 1972 The Arizona Board of Regents. Reprinted by permission of the University of Arizona Press)

Yellow Nutsedge (*Cyperus esculentus*) [P|EN]

Yellow Nutsedge is a persistent opportunist on the margins of my lawns and gardens. A North American native, the plant is now present in Australia, the Mediterranean region, South Africa, Yemen, Zambia, and probably elsewhere. Mowing or hoeing can control it (Fig. 9.133).

Fig. 9.133 Yellow Nutsedge (photo © Blahedo)

9.12 OLEASTER Family: ELAEAGNACEAE (Fig. 9.134)

Fig. 9.134 Russian Olive (*Elaeagnus angustifolia*)—ELAEAGNACEAE. Shrub or small tree to 10 m. Young twigs covered by silver scales become shiny brown with age. **Leaves**: Alternate, dark green with silvery scales, 4–10 cm long, 12–24 mm wide, smooth edges, and pointed tips. **Flowers**: Fragrant, yellow, single or in clusters of 2–3 near leaf axils; silver to white outside, yellow inside a tubular four-lobed calyx. **Fruit**: Short stalked with soft pulp surrounding a yellow 10–15 mm long achene covered with silvery-white scales (Fig. 9.134 painting, P.J. Redouté in Duhamel, 1804)

Russian Olive (*Elaeagnus angustifolia*) [P|DI]

With the Cold War at full intensity, Robert Austin in the Geography Department at the University of Utah became curious about Russian Olive. To see if the plant might have changed since its introduction from Asia to the USA, he requested information from herbaria throughout Russia, the plant's homeland and America's sworn enemy. He did not expect a response considering the strained relations between our countries. However, botanists at every herbaria sent photographs or negatives of their collections. Some even sent pressed plants. Espionage concerns had blocked nothing. The tree appeared unchanged.

Russian Olive grows well in cold dry regions. The USDA introduced it during the Dust Bowl of the 1930s for windbreaks, snowbreaks, soil stabilization, wildlife habitat, and pollen for honeybees. People also planted the tree for the beauty of its silver-green leaves and its bright yellow sweet-scented flowers. Later the species' invasive tendencies became obvious as it spread through the Great Plains and along waterways in the West.

The one tree at CWF bears heavy loads of sweet-scented flowers and fruit, but I have found no seedlings. Over the years, the tree has drooped, and it is shorter now than when we met 20 years ago (Fig. 9.135).

Fig. 9.135 Russian Olive. The one tree at CWF (photo © GR)

9.13 Spurge Family: EUPHORBIACEAE (Fig. 9.136)

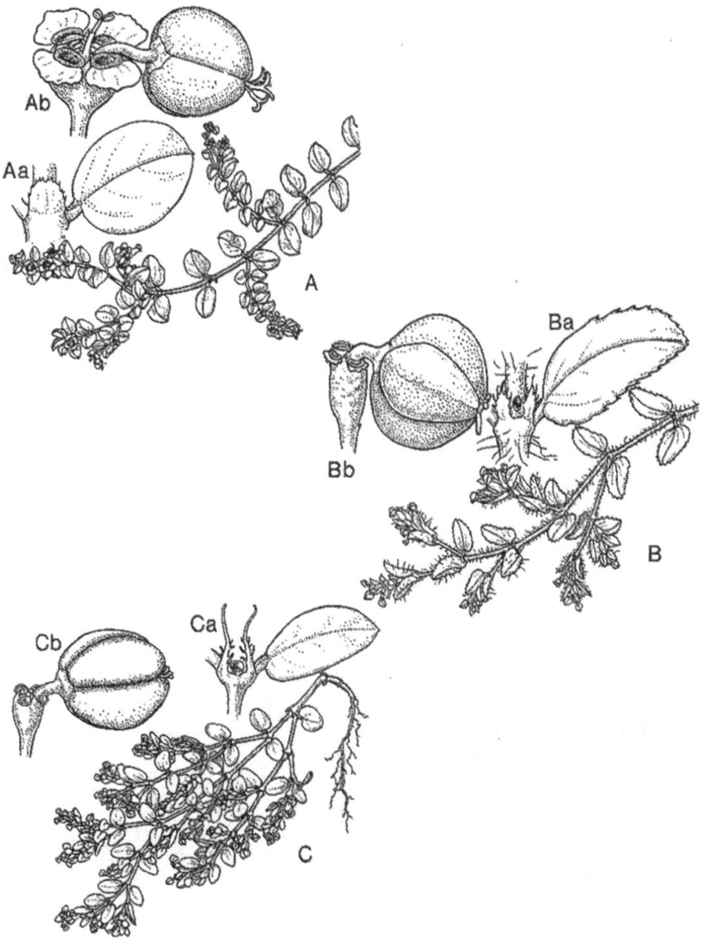

Fig. 9.136 Whitemargin Spurge (*Euphorbia albomarginata*) [P|N]. **A** in the drawing. Prostrate hairless perennial reproduces from seed and roots forming mats to 40 in. across. **Leaves**: Opposite, smooth edged often with a red spot and trimmed in white. Two stipules unite to form a white sail at leaf-stalk base. **Flowers**: Cyathia (involucre cup containing reduced stamens and single stigma) solitary in leaf axils. Involucres bell-shaped, hairless, 1 mm high ringed by four green-yellow to red oblong glands with a white to pink petaloid appendage often with a toothed edge; 15–30 staminate flowers. **Seeds**: Three-lobed hairless seedpods contain three whitish seeds 1–2 mm long. **Aa**. Leaf with white membranous scale at stalk base. **Ab**. Seedpod with four petaloid appendages and four involucre glands (Fig. 9.136 from *An Illustrated Guide to Desert Weeds*, by Kittie F. Parker. © 1972 The Arizona Board of Regents. Reprinted by permission of the University of Arizona Press). Sawtooth Spurge (*Euphorbia serrula*) [A|N]. **B** in the drawing. Prostrate to upright annual with *hairy stems*. **Leaves**: Opposite, *serrulate (saw toothed) edges* and uneven bases. Sometimes red along midvein. **Flowers**: Cyathia solitary or clustered in the nodes near branch tips. Involucres are cone-shaped to bell-shaped, 1 mm high, and almost hairless, with 4 greenish-yellow glands around the edge, each with a white to pink petaloid appendages wider than the gland; 7–15 staminate flowers. **Seeds**: Smooth capsules 2 mm long contain a smooth seed. **Ba**. Sawtoothed leaf with two tiny stipules at base. **Bb**. Seedpod with small glands and petaloid appendages. Littleleaf Spurge (*Euphorbia micromera*) [A|N]. **C** in the drawing. Prostrate annual. **Leaves**: Opposite, short stalked with even bases, less than 5 mm long with smooth edges. **Flowers**: Less than 1 mm, petaloid appendages absent, red to pink with 2–5 staminate flowers in five clusters around one pistillate flower with 3 styles and 3 ovary chambers. **Seeds**: Capsules 1–2 mm long with 1–2 mm long white to brown seeds. **Ca**. Leaf with two triangular stipules. **Cb**. Seedpod showing lack of involucral petaloid appendages (Fig. 9.136).

Spurges (*Euphorbia* spp.)

These three Spurge species along with Groundfig Spurge (next) grow in many places in CWF. Littleleaf was and is the most abundant around my home and was thick in the front lawn. We pulled it and now it grows in the gravel we spread over the driveway. The spurges produce no thorns or annoying seeds, deer browse them, so we leave them alone (Fig. 9.137).

Fig. 9.137 Littleleaf Spurge with mature seeds (photo © Stan Shebs)

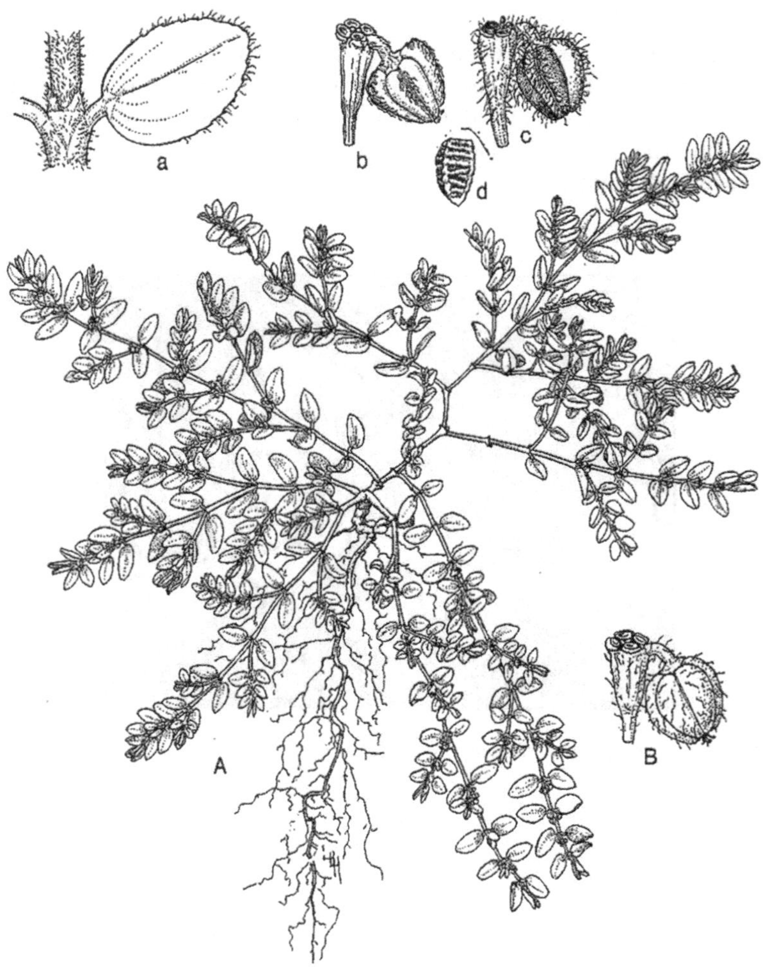

Fig. 9.138 Groundfig Spurge (*Euphorbia prostrata*)—EUPHORBIACEAE. Mat forming annual partially erect where competing for light. Pink to red stems have short spreading or curved hairs. **Leaves**: Opposite green or purplish leaves hairless above, sparsely hairy below, 3–8 mm long. Two tiny scalelike stipules may be present above the base of the leaf stalks. **Flowers**: Tiny pinkish flowers composed of stamens and pistils in small clusters. Three-lobed seedpods are less than 2 mm long, hairy on the three ribs. **Seeds**: Oblong, 1 mm long with four angles and 6–8 sharp wrinkles on each face. **A**. Prostrate plant. **a**. Leaf with tiny stipules near stalk base (not always present). **b**. Seedpod with ridge hairs lying on surface. **c**. Seedpod with erect ridge hairs. **d**. Groovy little seed. **B**. *Euphorbia supina* seedpod (Fig. 9.138 from *An Illustrated Guide to Desert Weeds*, by Kittie F. Parker. © 1972 The Arizona Board of Regents. Reprinted by permission of the University of Arizona Press)

Groundfig Spurge (*Euphorbia prostrata*) [P|T]

Though a tropical family, the spurges occur in many temperate subtropical locations like CWF. A Caribbean native, Groundfig Spurge is one of those few New World species that has established in the Old. In fact, the plant has such a long Old World history indigenous people have had time to learn several medicinal applications of the plant. In CWF, the plant competes with grasses, invading lawns and pastures where it grows below the reach of mower blades and is difficult to eradicate. Spend a few pleasant hours sitting in the grass pulling spurge and soon it will be gone.

An unusual characteristic of the CWF spurges (Groundfig and the previous three) is their opposite leaf pattern. Most members of the spurge family have alternate, not opposite leaves (Fig. 9.139).

Fig. 9.139 Groundfig Spurge (photo © Toony)

9.14 Pea (Legume) Family: FABACEAE (Fig. 9.140)

Fig. 9.140 Camelthorn (*Alhagi maurorum*)—FABACEAE. Matures at 18–48 in. tall. Reproduces from seed and from its roots and extensive rhizomes. Green stems bear green, yellow tipped spines 6–44 mm long. **Leaves**: Alternate 3–13 mm wide and 6–32 mm long. **Flowers**: Small 10 mm pea flowers are pink-purple to maroon along spine-tipped branches. **Seeds**: Like other peas, the seed-pod indents to show each of its 1–9 seeds. **a**. Pod normal size. **b**. Enlarged flower. **c**. Enlarged pod. **d**. Enlarged seed (Fig. 9.140 from *An Illustrated Guide to Desert Weeds*, by Kittie F. Parker. © 1972 The Arizona Board of Regents. Reprinted by permission of the University of Arizona Press)

Camelthorn (*Alhagi maurorum*) [P|FIPTZ2]

Camelthorn reminds me of childhood on another desert river, the Little Colorado in Northeastern Arizona. Across the street from my "country" home was a field of Rubber Rabbitbrush (*Chrysothamnus*), Fourwing Saltbush (Atriplex), and other shrubs and forbs. Beyond that, a broad Saltcedar (*Tamarix*) grove bordered the River. Camelthorn grew all along the roads and irrigation ditches, in our yard, and in patches almost everywhere else. Sharp thorns piercing our denim pants taught us to avoid any territory the plant had claimed.

This Eurasian perennial develops root systems to 6 ft. and can send up new shoots 20 ft. from parent plants. The roots will keep spreading until they meet bedrock or water barriers.

Though the thorny plant can be an impediment to travel on foot, it does enrich the sites it colonizes by accumulating nitrogen in root nodules and by moderating salinity as its roots decay.

Camelthorn has reached the area surrounding CWF. Once spotted here, I will go find a shovel (Fig. 9.141).

Fig. 9.141 Camelthorn flower (photo © Eitan F)

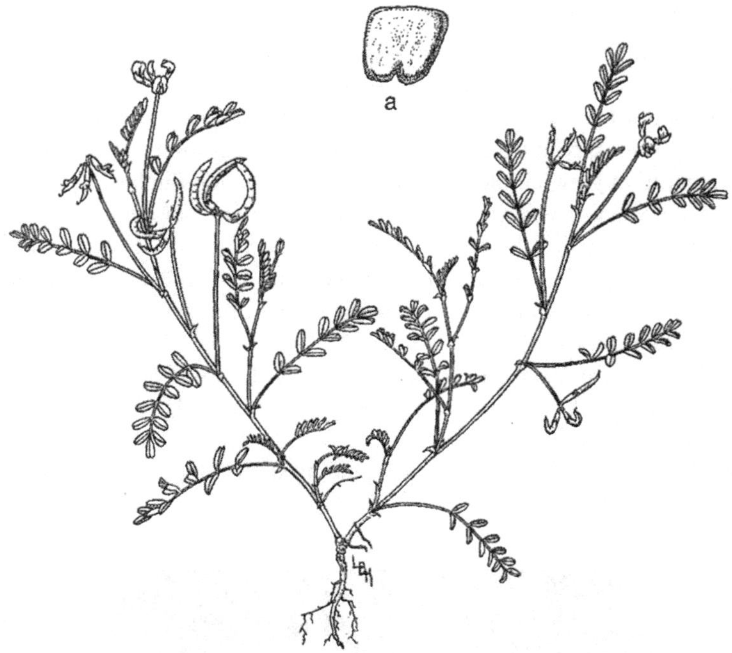

Fig. 9.142 Nuttall Milkvetch (*Astragalus nuttallianus*)—FABACEAE. Gray, slender annual with gray hair flat on surfaces, reproduces only from seed. Less than 1 ft. high spreads 6–40 in. **Leaves**: Three to six (sometimes eight) pairs of small leaflets with rounded, pointed, or notched tips. **Flowers**: Small, pale purple to almost white turn blue when dried. Two to six clusters at the ends of stiff stalks 1–3 in. long arising from leaf axils. **Fruit**: Flat pea pods 12–20 mm long and 2–3 mm wide curve upward. Pods remain on plants until they turn black. **a**. Seeds are 2 mm long, flat with a deep notch on one side. **a**. Seed (Fig. 9.142 from *An Illustrated Guide to Desert Weeds*, by Kittie F. Parker. © 1972 The Arizona Board of Regents. Reprinted by permission of the University of Arizona Press)

Nuttall Milkvetch (*Astragalus nuttallianus*) [A|N]

Common weed of disturbed sites in CWF and across the region. The plants will colonize lawns and gardens, but its small thin stems and leaves are no match for mower blades or springtime pulling. Like other legumes, Nuttall Milkvetch is toxic to livestock and its roots accumulate nitrogen. The nitrogen is a valuable nutrient used by other plants. Not a bad weed to have in some lawns; it is probably healthier than artificial fertilizers (Fig. 9.143).

Fig. 9.143 Nuttall Milkvetch (photo © Kaldari)

Fig. 9.144 Wooton Loco (*Astragalus allochorous var. playanusi*)—FABACEAE. Spreading annual reproducing from seed. Sprawling branches less than 12 in. long. **Leaves**: Pinnately compound, divided into 4–11 pairs of small gray leaflets 2–3 mm long. **Flowers**: Purple, 5–10, 6–8 mm long on stalks 2–7 cm long arising in leaf axils. Fruit: Thin, papery pods 7–28 mm long, straight on one edge. **Seeds**: Brown, flat, about 3 mm long with a notched edge. **a**. Seed. Halfmoon Loco (*Astragalus allochrous*)—FABACEAE. Erect to partially erect perennial very similar to Wooton Loco, but with longer branches (1–2 ft), more leaflets (11–19), and larger (>28 mm) seedpods (Fig. 9.144 from *An Illustrated Guide to Desert Weeds*, by Kittie F. Parker. © 1972 The Arizona Board of Regents. Reprinted by permission of the University of Arizona Press).

Wooton and Halfmoon Loco (*Astragalus allochorous*) [P|N]

Astragalus has more species and varieties worldwide than any other genus. There are over 200 in Arizona. Some are as weedy as Wooton and Halfmoon, but this pair will at least give you drawings and descriptions of two examples (Fig. 9.145).

Fig. 9.145 Wooton Loco (Siberian Elm seedling in the center) (photo © GR)

Fig. 9.146 White Sweetclover (*Melilotus albus*)—FABACEAE. Annual or biennial reproduces by seed, 1–6 ft. tall. Apart from flower color (white), White Sweetclover is very similar to Yellow Sweetclover, but *Yellow's stems are not as erect*. **Leaves**: Three leaflets on long stalks, similar to Yellow Sweetclover. **Flowers**: White, about 5 mm on spikelike stalks 2–8 in. long. **Seeds**: About 2 mm long in pods 2–3 mm long. **a**. Seed. **b**. Flower stalk with spikelike inflorescence (Fig. 9.146 from *An Illustrated Guide to Desert Weeds*, by Kittie F. Parker. © 1972 The Arizona Board of Regents. Reprinted by permission of the University of Arizona Press)

Wooton and Halfmoon Loco (*Astragalus allochorous*) [P|N]

Astragalus has more species and varieties worldwide than any other genus. There are over 200 in Arizona. Some are as weedy as Wooton and Halfmoon, but this pair will at least give you drawings and descriptions of two examples (Fig. 9.145).

Fig. 9.145 Wooton Loco (Siberian Elm seedling in the center) (photo © GR)

Fig. 9.146 White Sweetclover (*Melilotus albus*)—FABACEAE. Annual or biennial reproduces by seed, 1–6 ft. tall. Apart from flower color (white), White Sweetclover is very similar to Yellow Sweetclover, but *Yellow's stems are not as erect*. **Leaves**: Three leaflets on long stalks, similar to Yellow Sweetclover. **Flowers**: White, about 5 mm on spikelike stalks 2–8 in. long. **Seeds**: About 2 mm long in pods 2–3 mm long. **a**. Seed. **b**. Flower stalk with spikelike inflorescence (Fig. 9.146 from *An Illustrated Guide to Desert Weeds*, by Kittie F. Parker. © 1972 The Arizona Board of Regents. Reprinted by permission of the University of Arizona Press)

White Sweetclover (*Melilotus albus*) [A-B|D]

I did not find the origin of this species. It is present at mid and low latitudes on every continent except Antarctica. Bees and butterflies love this weed. If you want more of them in your yard, let a patch of White Sweetclover grow. Supplemental water may be necessary (Fig. 9.147).

Fig. 9.147 White Sweetclover (photo © Ivar Leidus)

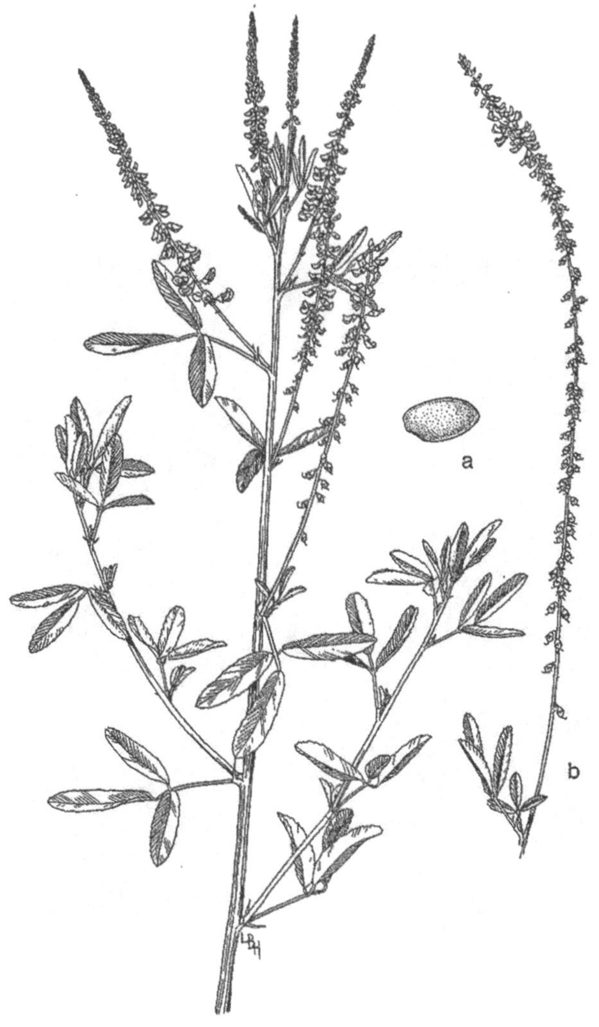

Fig. 9.148 Yellow Sweetclover (*Melilotus officinalis*)—LEGUMINOSAE. Annual or biennial reproduces by seed, 1–6 ft. tall. Very similar to White Sweetclover, but White's stems are more erect. **Leaves**: Three leaflets on long stalks, similar to White Sweetclover. **Flowers**: Yellow, about 5 mm on spikelike stalks 2–8 in. long. **Seeds**: About 2 mm long in pods 2–3 mm long. **a**. Seed. **b**. Flower stalk with spikelike inflorescence (Fig. 9.148 from *An Illustrated Guide to Desert Weeds*, by Kittie F. Parker. © 1972 The Arizona Board of Regents. Reprinted by permission of the University of Arizona Press)

Yellow Sweetclover (*Melilotus officinalis*) [P|DI]

Bees, birds, and butterflies love Yellow Sweetclover. Like other legumes, the plant stores nitrogen, and though it is an alien introduced species, it tends to be restricted to moist habitats across the deserts (Fig. 9.149).

Fig. 9.149 Yellow Sweetclover with Red-tailed Bumblebee (photo © Ivar Leidus)

Fig. 9.150 Annual Yellow Sweetclover (*Melilotus indicus*)—LEGUMINOSAE. This erect annual has spikelike inflorescences and small drooping flowers. **Leaves**: The compound leaves are 1–3 cm long and have three leaflets. They have small teeth above the middle and a dull red bar beside the midrib. *Their tips are indented or blunt.* **Seeds**: The round seedpod is about 2 mm long and holds one or two seeds. Seeds are eggshaped, 1–2 mm long, dark greenish brown, with rough surface. Flowers are 3–6 mm long, droop as they age. **a**. Seedling. **b**. Pealike flower. **c**. Seed (Fig. 9.150 from *An Illustrated Guide to Desert Weeds*, by Kittie F. Parker. © 1972 The Arizona Board of Regents. Reprinted by permission of the University of Arizona Press)

Annual Yellow Sweetclover (*Melilotus indicus*) [A|F]

This Mediterranean and Middle Eastern clover is near CWF, but I have not yet spotted it. Annual and sometimes biennial, the plant flowers from fall to spring in moist sites including lawns, gardens, streamsides, and farms.

Yellow Sweetclover feeds bees and other pollinators but is poisonous to some mammals. According to the Encyclopedia of Life, it can repel bed bugs and clear constipation. I have not checked the sources for these uses, but as the climate warms and humidity increases, repelling bedbugs might become a serious concern (Fig. 9.151).

Fig. 9.151 Annual Yellow Sweetclover. Tiny yellow flowers droop as they age (photo © Eigene Arbeit, selbst fotografiert)

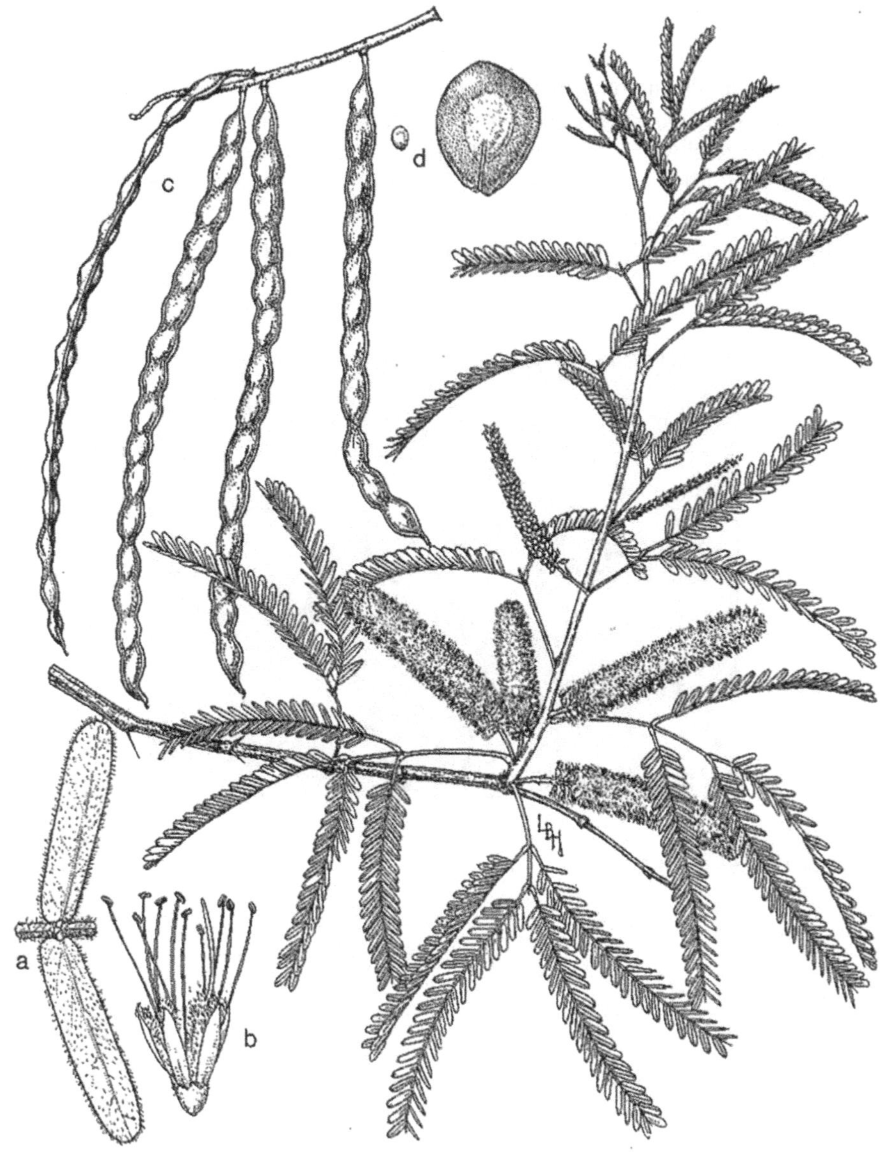

Fig. 9.152 Western Honey Mesquite (*Prosopis juliflora*)—FABACEAE. Deciduous shrub or small tree to 50 mm tall and trunk 1–4 ft. thick. Yellow spines in pairs 6–150 mm long. **Leaves**: Double pinnate compound 3–8 in. long divided into leaflets with 10–28 secondary leaflets 6–19 mm long. **Flowers**: Fragrant, yellow, crowded on stalked spikes 2–5 in. long. **Seeds**: Tan pods 3–8 in. long with sweet pulp and brown seeds. **a**. Pair of secondary leaflets. **b**. Flower. **c**. Seedpods. **d**. Seeds, one enlarged (Fig. 9.152 from *An Illustrated Guide to Desert Weeds*, by Kittie F. Parker. © 1972 The Arizona Board of Regents. Reprinted by permission of the University of Arizona Press)

Western Honey Mesquite (*Prosopis juliflora*) [P|DEIMN]

These small trees have slight presence in CWF and might have been abundant in the past. Cut for firewood and fence posts, mesquite numbers are probably much smaller than they were before Euro-American miners and farmers arrived. Mesquite is a weed of national significance in Australia and a serious problem in many other arid locations here.

Where Mesquite is a native, wildlife species use its flowers, beans, and seedpods. If you have a tree in a corner of the yard, your wildlife viewing will improve (Fig. 9.153).

Fig. 9.153 Western Honey Mesquite tree optimum growth at site with high water table. Shrub at right is a clump of mesquite sprouts from the stump of a tree cut for posts or fuel (photo © GR)

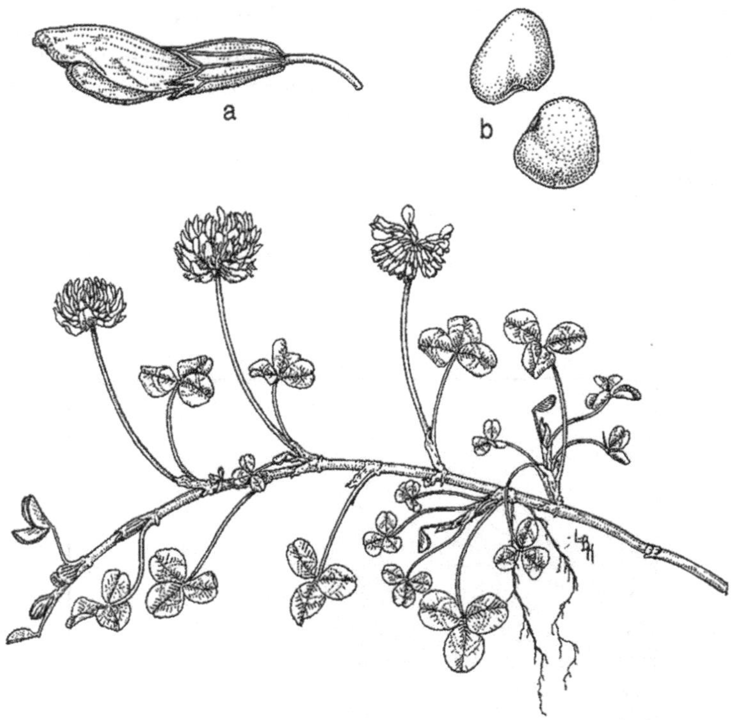

Fig. 9.154 White Clover (*Trifolium repens*)—LEGUMINOSAE. Hairless perennial with *creeping stems rooting at joints*. **Leaves**: Alternate, long erect stalks, three leaflets (or four every 10,000 leaves), notched at the apex. **Flowers**: Numerous tiny pink or white in globelike heads 1–3 cm wide. **Seeds**: Rounded, 2 mm long in pods. **a**. Flower. **b**. Seeds (Fig. 9.154 from *An Illustrated Guide to Desert Weeds*, by Kittie F. Parker. © 1972 The Arizona Board of Regents. Reprinted by permission of the University of Arizona Press)

White Clover (*Trifolium repens*) [P|D]

This interesting weed might confer good luck when it produces an extra leaflet, but more interesting is the plant's ability to flee its competitors. Only about one in 10,000 White Clover leaves will bear an extra leaflet, but all the plants you see are mobile across generations. White clover requires a moist site if it is to grow in CWF. You might find a small patch of clover shoots in your lawn. If you mark the boundary of the patch with small stakes, perhaps nails with a few inches of flagging, you can watch the patch roam about your lawn. John L. Harper described this behavior as the guerilla strategy wherein a plant sends up leaves and flowers from one area of shoots and lets other areas die. Thus, the plant avoids competition by shifting its position like a guerilla army.

White Clover attracts bees, butterflies, grasshoppers, moths, mice, javelina, and deer. I keep a patch in the yard for visitors and residents. Mowing the margins prevents spreading (Fig. 9.155).

Fig. 9.155 White Clover with a Common Checkered Skipper (*Pyrgus communis*). This butterfly prefers mallow flowers, but clover is good too (photo © GR)

9.15 Geranium Family: GERANIACEAE (Fig. 9.156)

Fig. 9.156 Redstem Filaree (*Erodium cicutarium*)—GERANIACEAE. Dark green winter annual or biennial reproducing from seed. Branching stems 3–24 in. long erect or prostrate with tips ascending. **Leaves**: Alternate, hairy, 1–10 cm long with 3–7 pairs of stalkless leaflets divided into fine segments; *first leaves form distinctive rosette*. **Flowers**: In umbrellalike clusters at ends of stalks arising from leaf axils. Five rose-purple petals are 6 mm or less long. **Fruits**: Needlelike, split into five one-seeded very hard fruits each with a *1–2 in. long tail that forms a tight corkscrew when dry*. Tails uncoil when wet and screw seeds into hardest soil. **a**. Basal rosette of young plant. **b**. Flower. **c**. Fruit with uncoiling tail (Fig. 9.156 from *An Illustrated Guide to Desert Weeds*, by Kittie F. Parker. © 1972 The Arizona Board of Regents. Reprinted by permission of the University of Arizona Press)

Redstem Filaree (*Erodium cicutarium*) [A|MI]

Also known as Heron's Bill and Stork's Bill, this is one of the most successful alien species in the Western United States. It grows in the warm Sonoran Desert south of CWF and the cold Great Basin to the north. Following wet winters, it produces seeds that burrow into the soil and germinate the next spring. The mature plants dry out by early summer, burn readily, and help create the weed vegetation replacing ecosystems worldwide (Fig. 9.157).

Fig. 9.157 Redstem Filaree by Otto Wilhelm Thomé, Flora von Deutschland

9.16 Watermilfoil Family: HALORAGACEAE (Fig. 9.158)

Fig. 9.158 Eurasian Watermilfoil (*Myriophyllum spicatum*)—HALORAGACEAE. Perennial aquatic herb 3–30 ft. long. Branches at water surface forming dense mat. **Leaves**: Whorls of four or five, pinnate with 12–20 segments per side. Uppermost leaves often squared at tip. **Flowers**: Green, born in terminal spike 5–10 cm long above water surface. Male flowers, present near the tip, have eight stamens. **Fruit**: Lobed nut-like cluster 2–3 mm long separate into four segments. **Winter buds**: Absent. **Insets**: **a**. Growth form. **b**. Whorl of leaves (x2). **c**. Flower spike with pistillate flowers below and staminate flowers above (x4). **d**. Immature fruit (x4). **e**. Mature fruit (x4) (Fig. 9.158 Drawing by Regina O. Hughes, U.S. Geological Survey, public domain)

Eurasian Watermilfoil (*Myriophyllum spicatum*) [P|FI]

This common fish-tank plant is present in local streams, lakes, and ponds worldwide. It is a problem when runoff from farms contains nitrogen and phosphorous fertilizers. The artificial enrichment permits the plant to expand in the spring. Together with floating algae, it can become so thick it blocks the light native plants need. This reduces native fish populations and forces waterfowl to move away to find more nutritious native plants. Eurasian Watermilfoil is spreading across North America. I have not seen the plant in CWF, but I expect it will arrive (Fig. 9.159).

Fig. 9.159 Watermilfoil (painting by Kurt Stüber—Pub Domain)

Fig. 9.160 Parrot's Feather (*Myriophyllum aquaticum*)—HALORAGACEAE. Aquatic, very leafy, stout stems can extend 1 ft. above water surface. **Leaves**: Whorled 2–5 cm long with 10–18 segments on each side, stalks 5–7 mm long. **Flowers**: In leaf axils and about 2 mm long. Only pistillate white flowers present in North America (Fig. 9.160 photo © André Karwath)

Parrot's Feather (*Myriophyllum aquaticum*) [P|FI]

Parrot's Feather is native to the Amazon River in South America, but today it has global distribution—cosmopolitan the botanists say. In North America, the plants are all female, reproducing from fragments broken off rooted plants. The weed invades water with high nutrient content, a characteristic of modern farm-polluted streams and lakes. It can grow so thick it blocks sunlight from native species and when those die, the fish depending on them die.

I have not seen the plant in the Agua Fria River or the River's floodplain ponds. However, pet shops often place Parrot's Feather in fish tanks. From there, a viable fragment will eventually make it into the river.

Neither physical nor chemical controls are effective with Parrot's Feather. Great Britain banned sale of the plant and other country bans will follow (Fig. 9.161).

Fig. 9.161 Parrot's Feather (photo © Harry Rose)

9.17 Iris Family: IRIDACEAE (Fig. 9.162)

Fig. 9.162 Yellow Flag (*Iris pseudacorus*)—IRIDACEAE. This Old World native forms clumps of stiff leaves 1.5 m tall with scattered yellow flowers 7 cm wide. Rhizomes are pink, branching, 2–3 cm in diameter. Seeds are flat, D-shaped, 6–7 mm wide, lustrous with shallow pits. Flowers appear April–June (Fig. 9.162 photo © GR)

Yellow Flag (*Iris pseudacorus*) [P|DIT]

Yellow Flag colonizes the borders of desert streams and ponds at middle latitudes. Nurseries distribute the plant despite its invasive tendencies. Once it establishes, Yellow Flag's robust roots and rhizomes make eradication of large stands difficult to impossible. Like many other weeds, it is important to remove this one when first noticed (Fig. 9.163).

Fig. 9.163 Yellow Flag at CWF (photo © GR)

9.18 Mint Family: LAMIACEAE (Fig. 9.164)

Fig. 9.164 Henbit (*Lamium amplexicaule*)—LAMIACEAE. Annual or biennial reproducing from seeds, stems rooting at lower joints, sometimes from rhizomes. *Square stems* 4–16 in. tall branch from the base. **Leaves**: Widely spaced, opposite, less than 1 in. long, hairy, with shallow lobes. Lower leaves on slender stalks, upper leaves broader, stalkless, and clasping. **Flowers**: Stalkless, born in upper leaf axils, pink or purple, tubular 8–17 mm long with two lips. The upper lip has spots, the lower is hairy on its back. **Seeds**: The yellow hairy calyx has five sharp bristlelike teeth. It remains on the plant and encloses a four-lobed fruit. The calyx separates into four triangular nutlets about 2 mm long. **a**. Flower. **b**. Seedlike nutlets (Fig. 9.164 from *An Illustrated Guide to Desert Weeds*, by Kittie F. Parker. © 1972 The Arizona Board of Regents. Reprinted by permission of the University of Arizona Press)

Henbit (*Lamium amplexicaule*) [A-B|E]

This Eurasian annual is present throughout the New World and is doing well around my house where it is spreading in the dry gravel between the bird garden and tractor shed. Of all the weeds that pop up there, it is the least offensive. Others have painful seeds or grow into giants that obstruct movement. Henbit works away at covering the gravel without causing any pain.

Henbit blooms from late February through the summer. The early start lets it make seeds before the spurges and taller plants shade it later in the spring. The flowers are nice, but too small to merit a D (Fig. 9.165).

Fig. 9.165 Henbit flowers (photo © Patrick Alexander)

Fig. 9.166 Horehound (*Marrubium vulgare*)—LAMIACEAE. Dense, woolly stems branching from the base grow 9–40 in. tall. Upper portions of stems are square. **Leaves**: Green above, white woolly underneath, opposite, corrugated, round with rounded teeth, 1–2 in. long including stalks. **Flowers**: Stalkless, white, tubular 6–8 mm long in dense clusters around the stem. The clusters are at the ends of all branches and often extending over one foot down the stem. The persistent calyx has hooked spines. **Seeds**: The calyx contains four nutlets each holding one eggshaped dark brown seed 2 mm long. **a**. Calyx with ten hooked spines. **b**. Views of the seedlike nutlet (Fig. 9.166 from *An Illustrated Guide to Desert Weeds*, by Kittie F. Parker. © 1972 The Arizona Board of Regents. Reprinted by permission of the University of Arizona Press)

Horehound (*Marrubium vulgare*) [P|EIP]

Horehound is a perennial alien abundant in CWF. It thrives in the warm arid climate of the Western United States and in similar environments worldwide. Hard candy and cough drops are its principal uses. Though Horehound use continues, it is a good example of the many plants that become problems when they escape cultivation.

Unlike the incoming tide of new weeds, Horehound has been here long enough to reach all the suitable sites in the region. When I moved to CWF in 1997, the plant had taken over the pastures and pond banks. I tried eradicating it by mowing it and spraying it with herbicides. Nothing worked, so I hired local kids to help me pull the plants. We pulled and disposed of mountains of Horehound. After that, mowing and occasional pulling along the pond banks was been effective.

Horehound disperses in animal fur. It has hooked seeds that bind fur into dense mats. To relieve your pets of their matted fur-pads, get out the scissors (Fig. 9.167).

Fig. 9.167 Horehound flowers (photo © GR)

9.19 Loosestrife Family: LYTHRACEAE (Fig. 9.168)

Fig. 9.168 Purple Loosestrife (*Lythrum salicaria*)—LYTHRACEAE. Perennial 50–150 cm tall. Forms many stemmed clones from a single woody root mass. **Leaves**: Opposite or whorls of three, stalkless, pointed, more than 5 cm long (on basal stem at lower left of drawing). **Flower**: Purple, crowded on spike 10–40 cm long, sepals and petals fused into slender tube. Petals wrinkled and separate for half their length of about 1 cm. **Fruit**: Many seeded two-chambered capsule enclosed by the sepal-petal tube (lower center of drawing) (Fig. 9.168 Drawing: Otto Wilhelm Thomé, 1885)

Purple Loosestrife (*Lythrum salicaria*) [P|FM]

According to contributors to the IUCN Global Invasive Species Database, this Eurasian weed is one of the 100 world's worst invasive species. It prefers moist sites in which it spreads by root and seed choking out native plants. Some roadside ditches in the region might give it access to CWF, but the most likely entrance is the Agua Fria River.

The plant supplies nectar and pollen to many insects. However, the threat it poses to native vegetation merits concentrated eradication efforts. Biological control using insects that feed on the plant has been successful. For instance, root feeding *Hylbius transversovittatus* and flower ovary feeding *Nanophyes marmoratus* weaken and restrain the weed's growth and spread (Fig. 9.169).

Fig. 9.169 Purple Loosestrife on the edge of a pond (photo © H. Zell)

9.20 Hibiscus Family: MALVACEAE (Fig. 9.170)

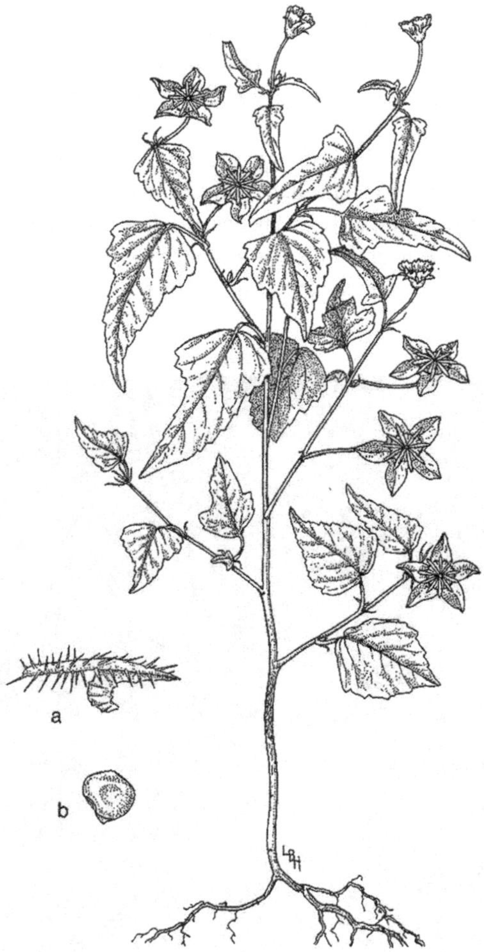

Fig. 9.170 Spurred Anoda (*Anoda cristata*)—MALVACEAE. Annual reproducing by seed. **Leaves**: Alternate triangular, 4–8 cm long with shallow lobes, toothed margins, and often with a red stripe down the center. **Flowers**: Pale purple, 2–3 cm long. **Fruit**: Flattened disk surrounded by five-lobed calyx. The disk has 9–20 sections that separate forming seedlike pods with dark hardened spur 2 mm long and covered with yellow hairs. **Seeds:** dark gray, wedgeshaped about 3 mm long. **a**. Spurred seedpod. **b**. Seed (Fig. 9.170 from *An Illustrated Guide to Desert Weeds*, by Kittie F. Parker. © 1972 The Arizona Board of Regents. Reprinted by permission of the University of Arizona Press)

Spurred Anoda (*Anoda cristata*) [A|IN]

Common around CWF, this annual weed has distinctive seedpods and sharp triangular leaves often with a central red stripe. It is notable for being an American native succeeding in the Old World. It is now present in Europe, Africa, Japan, and Australia.

You can control small infestations of Spurred Anoda, but chemical-free safe control when plants have infested a crop requires hoeing, mulching, and crop rotation (Fig. 9.171).

Fig. 9.171 Spurred Anoda fruit (photo © GR)

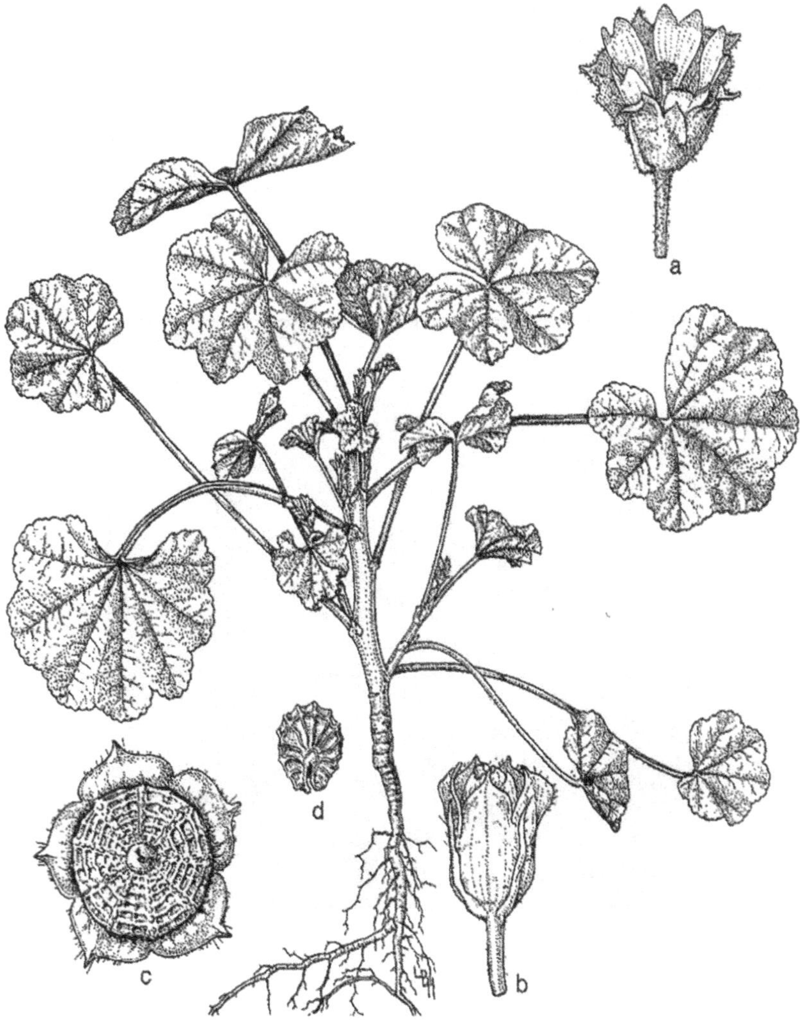

Fig. 9.172 Little Mallow (*Malva parviflora*)—MALVACEAE. Spreading annual 1–3 ft. tall. **Leaves**: Alternate, circular, 2–5 in. wide, on stalks 4–10 in. long, with 5–7 shallow lobes, toothed edges, and a red spot near the base. **Flowers**: Pale blue or pink, 6 mm long, in clusters at base of leaf stalks. **Seeds**: Outer green sepals spread under the flower forming a round disk of 11–12 carpels. Each carpel contains one 2 mm long brown seed. **a**. Flower. **b**. Closed flower in cooler temperatures. **c**. Fruit surrounded by five sepals. **d**. Carpel containing one seed (Fig. 9.172 from *An Illustrated Guide to Desert Weeds*, by Kittie F. Parker. © 1972 The Arizona Board of Regents. Reprinted by permission of the University of Arizona Press)

Little Mallow (*Malva parviflora*) [A-B-P|M]

This weed is short-lived but it has strong roots that resist pulling. Mowing will keep the plants from spreading, but you have to pull them for complete eradication. Little Mallow has a long flowering season, and seeds mature quickly. So, if you are shooting for eradication, pull plants no later than the first flowers in spring (Fig. 9.173).

Keep a few plants around for their medicinal value. Dried plants have the most applications, but Moore says crushed fresh plants make great poultices for stings, inflammations, and pustular infections (Moore 2003, pp. 155–156).

Fig. 9.173 Little Mallow flowers (photo ©GR)

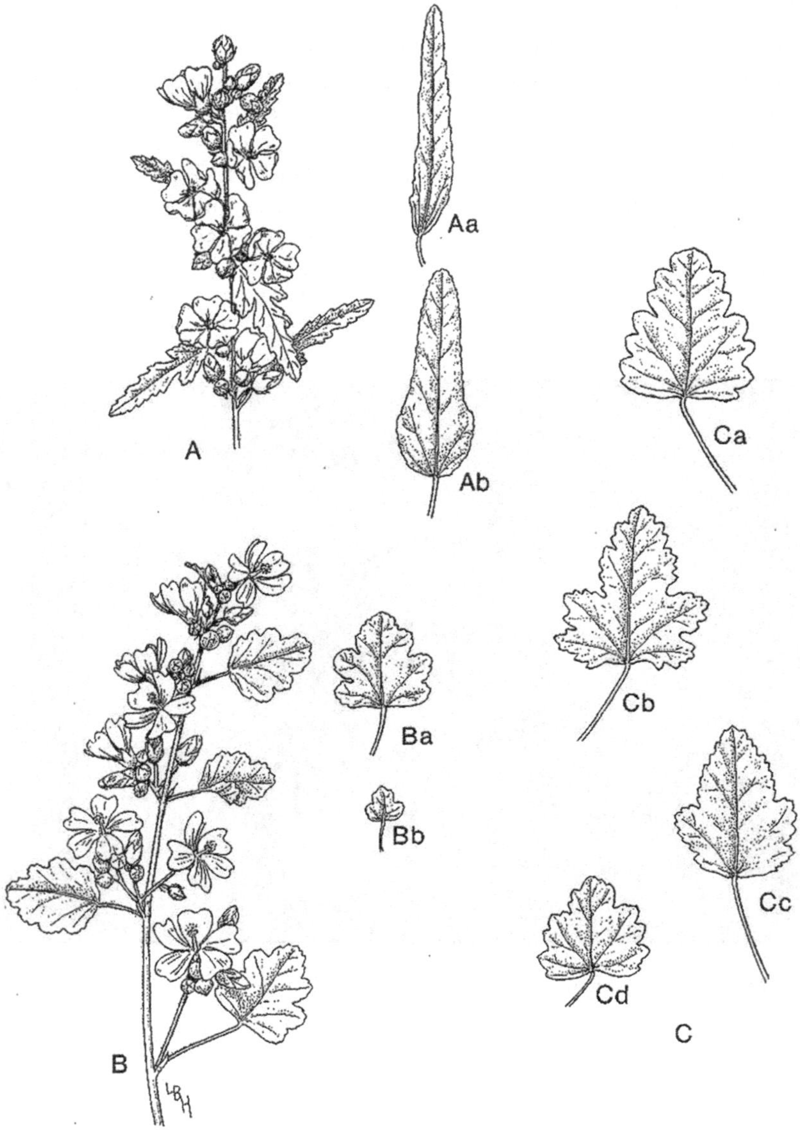

Fig. 9.174 Littleleaf Globemallow (*Sphaeralcea parvifolia*)—MALVACEAE. B above: Perennial shrub 20–100 cm tall with gray or white hair covering leaves and stems. New stems grow from thick woody root crown. **Leaves**: Alternate, 2–5 m long with shallow lobes and toothed margins. **Flowers**: Deep orange with five petals 8–18 mm long, born in short racemes or panicles. **Fruit**: Hemispherical with 9–12, 2–5 mm long carpels with two seeds. **Ba** and **Bb**: Other leaf types. Narrowleaf Globemallow (*Sphaeralcea angustifolia*) A above, a native weed is present near CW Caliche Mallow (*Sphaeralcea laxa*), **Ca, Cb, Cc, Cd** above, also native, might be present on the old volcanic soils covering the hill southwest from the Farm (Fig. 9.174 from *An Illustrated Guide to Desert Weeds*, by Kittie F. Parker. © 1972 The Arizona Board of Regents. Reprinted by permission of the University of Arizona Press).

Littleleaf Globemallow (*Sphaeralcea parvifolia*) [P|MN]

The Globemallows (*Sphaeralcea*) are taller and have larger flowers than Little Mallow (*Malva*). The plants are perennial, but mowing will control their spread. However, the flowers are nice and they attract bees and butterflies (Fig. 9.175).

Another reason to keep a few plants around the house is their medicinal value. Moore says crushed fresh leaves make superb poultices for stings, inflammations, and pustular infections (Moore 2003, pp. 155–156).

Fig. 9.175 Littleleaf Globemallow (photo © GR)

9.21 Unicorn-Plant Family: MARTYNIACEAE (Fig. 9.176)

Fig. 9.176 Devil's Claw (*Proboscidea parviflora*)—MARTYNIACEAE. To 2.5 m wide and 1 m tall. **Leaves** broad triangles to 25 cm long and 22 cm wide, margins entire or 3–7 shallow lobes. **Flowers** hidden among leaves, tubular corolla pink, purple, or white, 5–15 on racemes to 25 cm, pedicels 1–3 cm long fruit developing to 45 mm. **Fruit** green succulent capsule 5–7 cm long and 2 cm wide, glandular-pubescent when young, becoming woody when dry. Spectacular tough springy hooks 10–15 cm long. Hooks clasp paws, hooves, and shoes (Fig. 9.176 photo © GR)

Devil's Claw (*Proboscidea parviflora*) [A|DEN]

With its interesting seedpods and tolerance across the varying climates of all North American Deserts, Devil's Claw is a strong candidate for dispersal to Old World deserts. The capsules might be edible when immature, but I have not tried them (Figs. 9.176, 9.177, and 9.178).

Fig. 9.177 Devil's Claw fruit. Though sharp tipped, the horns achieve dispersal mainly by clasping the feet of passing animals (photo © GR)

Fig. 9.178 Devil's Claw flowers (photo © GR)

9.22 Mulberry Family: MORACEAE (Fig. 9.179)

Fig. 9.179 White Mulberry (*Morus alba*)—MORACEAE. Shrubs or trees to 15 m. Red or brown bark with thin red or yellow furrows and long narrow ridges. **Leaves**: Pointed, toothed, 6–10 cm long, and 3–6 cm wide, often with deep lobes. **Flowers**: Male (white or yellow-green) and female (green) in drooping catkins on same or different plants. Sepals green with red tips. **Fruit**: Female transform into compound fruits. **Seeds** in light brown achenes 2–3 mm in diameter (Fig. 9.179 drawing, Bailey, 1917)

White Mulberry (*Morus alba*) [P|DE]

Birds drop mulberry seeds everywhere in CWF. Trees grow on lawn and garden margins and other places where there is extra moisture. If you want to keep the tree, trim all but one central shoot. Keep this up, and within a few years, you will have a fine shade tree. If a seedling appears where you do not want a tree, do not wait too long to remove it. The longer you wait the more the roots develop and the more likely they will produce new shoots to replace any you cut (Fig. 9.180).

The trees produce all black or all white fruit. I cannot taste the difference, but one visitor said she thought black was best.

Fig. 9.180 White Mulberry. Young tree with a plethora of basal sprouts (photos © GR)

9.23 Four O'clock Family: NYCTAGINACEAE (Fig. 9.181)

Fig. 9.181 Red Spiderling (*Boerhavia coccinea*)—NYCTAGINACEAE. Perennial with stout base and leaves but with fine stems radiating outward up to 6 ft. from the thick woody root. Stems often have sticky yellow bands above and are hairy near the base. **Leaves**: Opposite, eggshaped, 1–4 in. long, tomentose to hairless. Similar to Coulter Spiderling (next). **Flowers**: Small, *brilliant magenta, umbrellalike clusters of 3–25 at tips of the many threadlike stems* forming at maturity. **Fruit**: Five ribs, about 3 mm long, hairy, sticky. **a**. Umbrellalike flower cluster. **b**. Fruit covered by glands emitting a sticky substance (Fig. 9.181 from *An Illustrated Guide to Desert Weeds*, by Kittie F. Parker. © 1972 The Arizona Board of Regents. Reprinted by permission of the University of Arizona Press)

Red Spiderling (*Boerhavia coccinea*) [P|DN]

More adventurous than Coulter Spiderling (next), Red Spiderling grows in many places around the world. Its origin is uncertain, but best guess is the Southwestern United States. The leaves are attractive, and the flower is brilliant (best observed with your loop). The allelopathic plants emit powerful chemicals that block other weeds and secure sites for future growth. Control spiderlings with an occasional mow job (Fig. 9.182).

Fig. 9.182 Red Spiderling flower cluster (photos © GR)

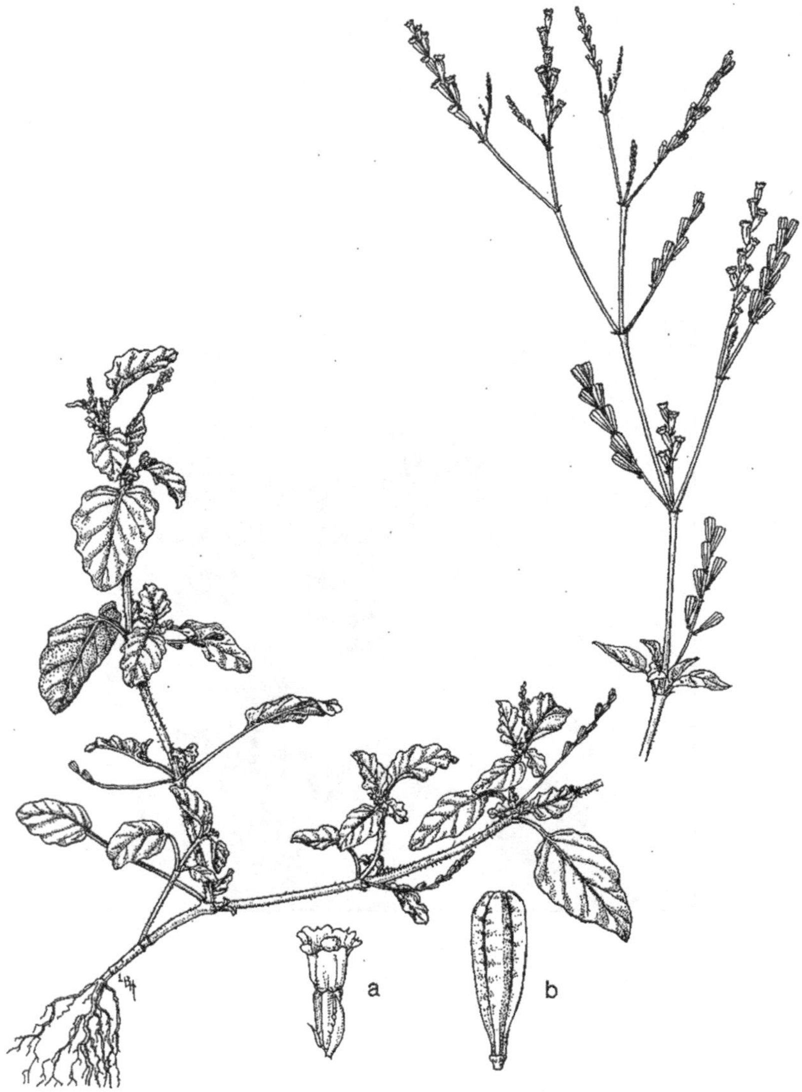

Fig. 9.183 Coulter Spiderling (*Boerhavia Coulteri*)—NYCTAGINACEAE. Annual reproducing by seed; erect 12–36 in. high or prostrate stems radiating outward 1–5 ft. and rising at the tips. Lower stem hairy, upper stem sometimes with sticky yellow bands at maturity. **Leaves**: Opposite, eggshaped, hairless except along stalks and sometimes edges, 1–4 in. long. Similar to Red Spiderling (cf.). **Flowers**: *Pale pink*, about 2 mm long, solitary or in clusters of two or three, *scattered along upper stems for 1–2 in*. **Fruits**: Wedgeshaped, hairless, about 3 mm long with five ridges. **a**. Flower. **b**. Fruit (Fig. 9.183 from *An Illustrated Guide to Desert Weeds*, by Kittie F. Parker. © 1972 The Arizona Board of Regents. Reprinted by permission of the University of Arizona Press)

Coulter Spiderling (*Boerhavia Coulteri*) [A|FN]

This native annual blooms from mid to late summer. Like Red Spiderling, the flowers are attractive, but they are so small (1–2 mm) you will need your hand lens to see them. I have not found this species in CWF, but it grows at similar elevations throughout Arizona, and is probably here now, or will be soon.

This is one of the least studied weeds in this book. There are photos and taxonomic descriptions on the Web, but I found nothing about the plant's value for wildlife or humans. If you find this plant interesting, you can contribute an article about it to the Encyclopedia of Life, and you could expand the stub article in Wikipedia (Fig. 9.184).

Fig. 9.184 Coulter Spiderling (photo © Stan Shebs)

9.24 Evening Primrose Family: ONAGRACEAE (Fig. 9.185)

Fig. 9.185 Velvety Gaura (*Oenothera curtiflora*)—ONAGRACEA. Annual 30–200 cm tall, covered by soft long hair, branches near the top. **Leaves**: Basal cluster, alternate along stems, lanceolate 2–13 cm long and 5–40 mm wide. Small teeth line edges near the tip. **Flowers**: Small, asymmetrical, pale pink or white, self-pollinating, in 10–30 cm long spikes at branch tips. The floral tube below the sepals is 1–5 mm long, four drooping sepals 2–4 mm long; four petals 2–3 mm long. Petals fade to dark pink or red. **Fruits**: Woody capsule 5–10 mm long with four angles, tapers to slender base. One-four oval seeds remain within the capsule (Fig. 9.185 photo © Mack Hitch)

Velvety Gaura (*Oenothera curtiflora*) [A|N]

This tall native has distinctive, disorganized flower spikes extending up to a foot from its shoot tops. My next-door neighbor called it fireweed and claimed it caused rashes. However, the plant's dense hair coat is harmless and indeed feels like velvet (Fig. 9.186).

Fig. 9.186 Enlarged section of Velvety Gaura flower spike (photo © Matt Lavin)

9.25 Woodsorrel Family: OXALIDACEAE (Fig. 9.187)

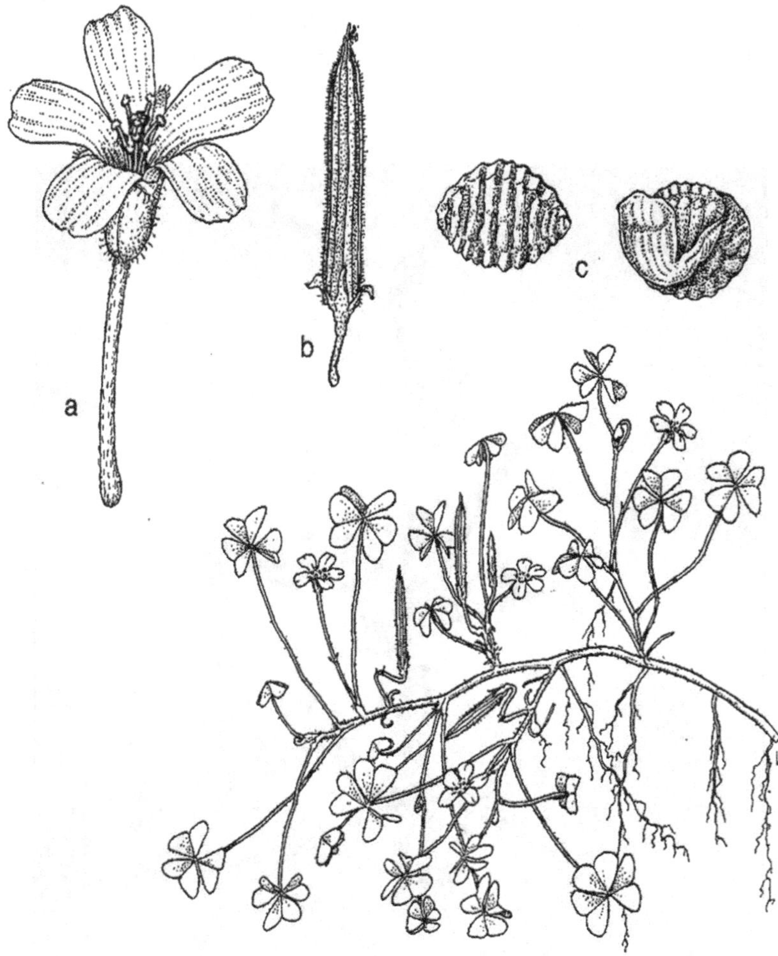

Fig. 9.187 Creeping Woodsorrel (*Oxalis corniculata*)—OXALIDACEAE. Prostrate with creeping stems 3–8 in. long. Reproduces from seed, stems rooting at joints, and from rhizomes. **Leaves**: Alternate leaves with *three heartshaped leaflets* at the tip of long leaf stalks. The green, purple, or bronze leaflets close and droop at night. **Flowers**: Five yellow petals 3–8 mm long. One to five clusters at the end of stalks rising from leaf axils. Similar to Bullhead (below). **Seeds**: Tan seedpods are erect on bent stalks. They are cylindrical, 5-sided, 8–25 mm long with pointed tips. When the 1 mm long seeds mature, the pod bursts open shooting the many tiny brown hopefuls up to 10 ft. **a**. Flower. **b**. Seedpod. **c**. Two views of seed (Fig. 9.187 from *An Illustrated Guide to Desert Weeds*, by Kittie F. Parker. © 1972 The Arizona Board of Regents. Reprinted by permission of the University of Arizona Press)

Creeping Woodsorrel (*Oxalis corniculata*) [P|DE]

We can live with this plant. A rare introduction from South America, Creeping Woodsorrel may have originated in the Andes Mountains of Ecuador. People have spread it around the globe; I once found it in Manhattan, NY growing beneath Yew shrubs in a small planter surrounded by 20 ft. of concrete. The plant colonizes exposed soil and is persistent on the margins of lawns. It is present in shady moist sites throughout CWF. Dispersal is by exploding seedpods besides the common vehicles. Like the spurges and Plantago, much of Creeping Woodsorrel's foliage is below the bite of mower blades. Woodsorrel is not very competitive, however, and does not force its way between lawn grasses. The flower color and size resemble Bullhead, but the leaves are different. So check, because if it is Bullhead you will want to pull the monster before it makes seeds.

The deer, gophers, and occasional rabbit visiting my yard like to nibble Woodsorrel foliage, and the Deer mice, White-footed mice, Mourning Dove, Juncos, Field Sparrows, Grasshopper Sparrows, Savannah Sparrows, and Tree Sparrows eat the seeds. Bees, butterflies, and flies go for the flower nectar and pollen. Caterpillars of at least one species of the Cyna (*Zizula*) butterflies prefer these plants.

Every bit of this weed is edible, but do not eat bales of it because it contains some oxalic acid. It has a tangy flavor strongest in the stems and ripening seedpods (Fig. 9.188).

Fig. 9.188 Creeping Woodsorrel flowers (photo © GR)

9.26 Poppy Family: PAPAVERACEAE (Fig. 9.189)

Fig. 9.189 Crested Pricklepoppy (*Argemone pleiacantha*)—PAPAVERACEAE. Broken leaves and stems exclude yellow sap. Large white **flowers** 2–3 in. wide with 4–6 petals and many lemon-yellow stamens. **Leaves** are alternate, lobed, clasping at the base, and 2–8 in. long. **a**. Crested Pricklepoppy flower bud with three spiny terminal horns. **b**. Seedpod, 2–4 cm long. **c**. Seed, 2–4 mm in diameter. **d**. Bluestem Pricklepoppy plants are very similar to Crested Pricklepoppy. Crested is bluer and its *flower buds have three terminal horns that lack spines except at base*. (Fig. 9.189 from *An Illustrated Guide to Desert Weeds*, by Kittie F. Parker. © 1972 The Arizona Board of Regents. Reprinted by permission of the University of Arizona Press)

Crested Pricklepoppy (*Argemone pleiacantha*) [A-B|MN]

Pricklepoppy is a native of Arizona and the southern and western Great Plains. You cannot miss them. They have large white flowers with bright lemon-yellow centers. Plants bloom from March to August and their fragrance attracts many ants, beetles, and butterflies (Fig. 9.190).

Though Pricklepoppy requires little moisture, it occupies depressions and roadside ditches where extra moisture collects. It is also common where heavy grazing eliminates grasses and other competitors and breaks up the soil surface. Why does this plant remain? Well, wait until you see it. It is very thorny; plus, if you break a branch, it bleeds a repugnant yellow sap that repels grazing animals. I like this comment by Michael More: "Often a symptom of fierce overgrazing, I have heard otherwise sentient, grizzled old ranchers fussily insist that the plant 'chokes out the grass,' while surveying their sad old pastureland covered only in plants that cattle can't eat. And hens mature into eggs" (Moore 2003, p. 210).

Fig. 9.190 Crested Pricklepoppy, CWF, Arizona (photo © GR)

Fig. 9.191 Golden Corydalis (*Corydalis aurea*)—PAPAVERACEAE. Low bluish green biennial reproducing by seeds. Stems contain watery juice. Branches often lay on the ground for a short distance and then grow upward. **Leaves**: Deep segmented lobes. **Flowers**: Bright yellow, 1–2 cm long with spur at the base and two sepals. **Seeds**: Bluegreen cylindrical seedpods 13–25 mm long curve downward. The plethora of shiny black seeds are about 2 mm long. **a** Flower with spur. **b**. Seed (Fig. 9.191 from *An Illustrated Guide to Desert Weeds*, by Kittie F. Parker. © 1972 The Arizona Board of Regents. Reprinted by permission of the University of Arizona Press)

Golden Corydalis (*Corydalis aurea*) [B|N]

This is a native biennial weed. Scrambled Eggs is another common name, but I am sticking with Golden Corydalis.

The weed is present throughout this region but I have not seen it in CWF—yet. There are two variants: *Aurea* and *occidentalis*. *Aurea's* fruit spreads or dangles and the flowers do not grow taller than the leaves. *Occidentalis's* fruit is erect and the flowers grow taller than the leaves (photo below). The weed grows in the altitude range from 2500 to 7500 ft. in moist riparian areas. In the past, botanists reported *Occidentalis* prefers warmer conditions and grows between 2500 ft. and 4000 ft. However, recent global warming has raised average temperatures almost 1.5 degrees Celsius. The adiabatic lapse rate being 1.5 degrees Celsius per 500 ft., conditions in CWF are becoming suitable for *Occidentalis*. Those same botanists found *Occidentalis* preferred dry soil, which should give it an advantage in CWF and all other drylands (Fig. 9.192).

Fig. 9.192 Golden Corydalis (photo © Jerry Friedman)

9.27 Plantain Family: PLANTAGINACEAE (Fig. 9.193)

Fig. 9.193 Buckhorn Plantain (*Plantago lanceolata*)—PLANTAGINACEAE. Small perennial 12–18 in. tall with thick fibrous roots; reproduces from seed. **Leaves**: Ribbed, 3–12 in. long, 6–25 mm wide with pointed tips; leaf axils often filled with dense brown hair. **Flowers**: Spikes 2–5 cm on stalks longer than the leaves. **Fruit**: Spherical, papery pods about 3 mm long, upper half falls off to release seeds. Seeds about 2 mm long, shiny, brown. **a**. Two views of boat-shaped seed (Fig. 9.193 from *An Illustrated Guide to Desert Weeds*, by Kittie F. Parker. © 1972 The Arizona Board of Regents. Reprinted by permission of the University of Arizona Press)

Buckhorn Plantain (*Plantago lanceolata*) [P|EFM]

Buckhorn Plantain's origin is not clear in the literature. It has a long history in both Eurasia and N. Africa, and pollen analysis shows it followed the spread of Neolithic farmers. Today it grows in temperate habitats worldwide. A vague memory tells me I have seen the plants somewhere around the farm, perhaps beside the river. I will look there again next spring. Like its big-leafed cousin (Broadleaf Plantain), Buckhorn Plantain prefers moist sites and colonizes roadsides and watered lawns and gardens. Native Americans and others use the plant to remedy headaches, dysentery, poisonous bites, stings, burns, blisters, sore eyes, bowel complaints, earaches, and hungry cows. Moore (2003) describes additional uses and provides details on collection and preparation.

Wind, beetles, and flies spread the sneeze-inducing pollen of Buckhorn Plantain throughout spring and summer (Fig. 9.194).

Fig. 9.194 Buckhorn Plantain (photo © Sanrise)

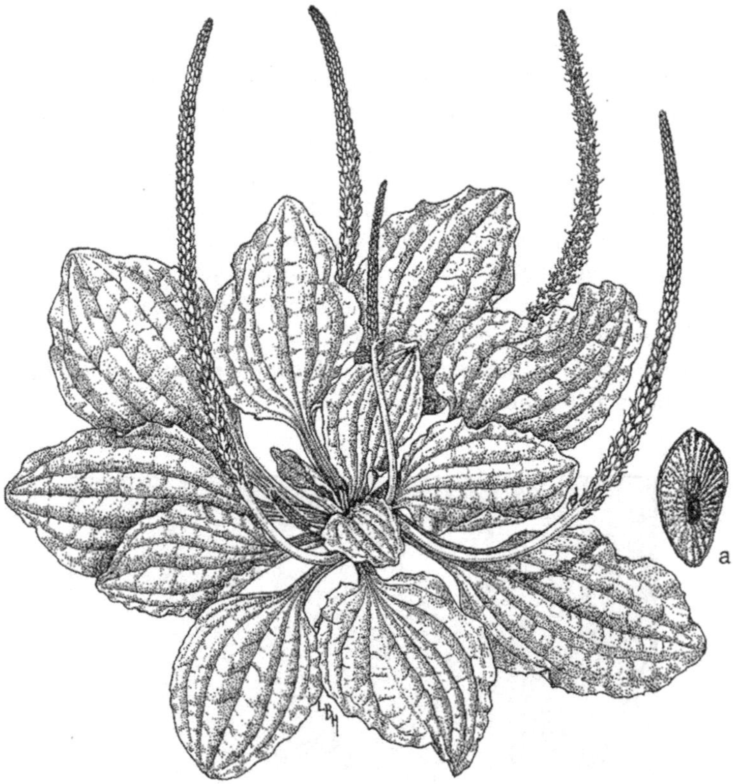

Fig. 9.195 Broadleaf Plantain (*Plantago Major*)—PLANTAGINACEAE. No aboveground stems. Leaves 3–8 in. on stalks 2–5 in. long and 2–4 in. wide, 5–7 ribs, and smooth wavy to toothy margins. Flower spikes 3–12 in. long. The small four-lobed flowers are thin, pale, dry, and persistent. Seedpods globular, 3 mm long contain 6–20 seeds 1–2 mm long (Fig. 9.195 from *An Illustrated Guide to Desert Weeds*, by Kittie F. Parker. © 1972 The Arizona Board of Regents. Reprinted by permission of the University of Arizona Press)

Broadleaf Plantain (*Plantago major*) [P|EM]

This Eurasian perennial appears in moist locations in CWF and in most of the world. It forms a strong root system, and though it can sprout from its roots, it reproduces mainly by seed. Broadleaf Plantain in lawns adds texture and diversity perfectionists detest. It is too flat to the ground to mow, so pulling or poisoning is the option (Fig. 9.196). The plant has beneficial uses ranging from salad greens to pain relief to cough medicine and many more. If you want to learn to use Plantago and other plants for natural medicine, read Michael Moore's *Medicinal Plants of the Mountain West* (Moore 2003).

Fig. 9.196 Photo © Jesse Taylor

Fig. 9.197 Yellow Toadflax (*Linaria vulgaris*)—PLANTAGINACEAE. Perennial 30–80 cm tall forms patches by sprouting from roots. **Leaves**: Alternate, pale green, 2–5 mm wide; *not clasping like Dalmatian Toadflax*. **Flowers**: Yellow with orange centers, 2–4 cm long including nectar spurs. **Fruit**: Round, 8–12 mm wide capsule containing numerous small brown winged seeds (Fig. 9.197 painting by Norman Criddle)

Yellow Toadflax (*Linaria vulgaris*) [P|IT]

A European native, Yellow Toadflax is a common garden flower across the USA. However, this weed is a serious threat to native vegetation. The plants produce extensive root systems from which new shoots arise when there is sufficient moisture for growth. The roots can lie dormant for up to 10 years. Once established in native vegetation, control becomes very difficult (Fig. 9.198).

Fig. 9.198 Yellow Toadflax (photo © Jeff Delong)

Fig. 9.199 Dalmatian Toadflax (*Linaria dalmatica*)—PLANTAGINACEAE. Three-foot tall perennial sprouts from a deep spreading root system. **Leaves**: Crowded on stem, ovate, sharp tips, and *clasping at the base*. **Flower**: Five-lobed, two lipped, bright yellow corolla 2–5 cm long with a spur at the base (Fig. 9.199 photo © Peganum)

Dalmatian Toadflax (*Linaria dalmatica*) [P|ITZ-2]

Many people have planted this alien perennial weed in their yards for the bright yellow snapdragon-like flowers. As time passes, however, the flowers magically appear everywhere in your yard, along the street, and in the neighbors' yards. This is not a problem in yards and gardens; yearly mowing suffices to maintain control.

However, such intense management is not practical across miles and miles of the pinyon-juniper woodlands and Ponderosa Pine forest where Dalmatian Toadflax has spread. There, the plant competes with and replaces native grasses and wildflowers. It spreads into areas where livestock have thinned out native plants and broken up the soil surface. Once established, Dalmatian Toadflax will not readily give up its place. Each plant produces up to 500,000 seeds that remain viable in the soil for 10 years or more. Roots remain viable for even longer. In the Coconino National Forest northeast of CWF, Dalmatian Toadflax has invaded more than 200,000 acres (Phillips and Crisp 2001).

Common Buckeye Caterpillars (*Junonia coenia*) eat Toadflax. Perhaps a swarm will thin out the plants out in the woods next spring. We can always use more butterflies (Fig. 9.200).

Fig. 9.200 Dalmatian Toadflax. Pods opening reveal seeds (photo © GR)

9.28 Grass Family: POACEAE (Fig. 9.201)

Fig. 9.201 Wild Oats (*Avena fatua*)—POACEAE. Annuals 8–160 cm tall. **Leaves**: Sheaths of basal leaves with scattered hair; upper sheaths hairless; ligules 4–6 mm long, pointed; blades 10–45 cm long and 3–15 mm wide. **Flowers**: Nodding, panicles 7–40 cm long and 5–20 cm wide. **Seeds**: Two or more spikes extend from each seed. (Fig. 9.201 painting by Norman Criddle in Clark and Fletcher 1909)

Wild Oats (*Avena fatua*) [A|EM]

You can often see big-seeded Wild Oats from the car window as you whiz along. Is this roadside vagrant the archetype for the wild behavior of youth? *Fatua* is Latin for foolish, insipid, or worthless. But what is it about this plant that caused Linnaeus to label it worthless? Its genus, *Avena* is Latin for oats and the seeds are large and edible; why is the plant a weed and not a commercial field crop? In fact, Wild Oats is in commercial oat fields around the world, but not by design. The plant invades *Avena sativa* (*sativa* means domestic or planted) crops where it reduces overall productivity by using water and soil nutrients to make more leaves but fewer seeds than *sativa*. I guess from the human point of view, this is foolish. However, Kane and Moore report above average medicinal uses for Wild Oats (Kane 2017; Moore 1989). So, not too worthless (Fig. 9.202).

Fig. 9.202 Wild Oat spikelets (photo © Alvesgaspar)

Fig. 9.203 Sixweeks Needle Grama (*Bouteloua aristidoides*)—POACEAE. Small short-lived annual 3–12 in. tall, reproducing by seeds. **Leaves**: Thin, 3–6 in. long and 2 mm wide; ligule hairy, small 1–2 mm, membranous and lacerate or ciliate. **Flowers**: 4–20 spikes with pointed bases and spreading tops, 1–2 cm long. Loosely attached along one side of the stem; each with three short bristles about 5 mm long. **Seed**: Brown, about 3 mm long. **a**. Two spikelets. **b**. Lowest spikelet. **c**. Grain (Fig. 9.203 from *An Illustrated Guide to Desert Weeds*, by Kittie F. Parker. © 1972 The Arizona Board of Regents. Reprinted by permission of the University of Arizona Press)

Sixweeks Needle Grama (*Bouteloua aristidoides*) [A|P]

This species has about the same distribution as Sixweeks Grama, but is the more abundant of the two at CWF (Fig. 9.204).

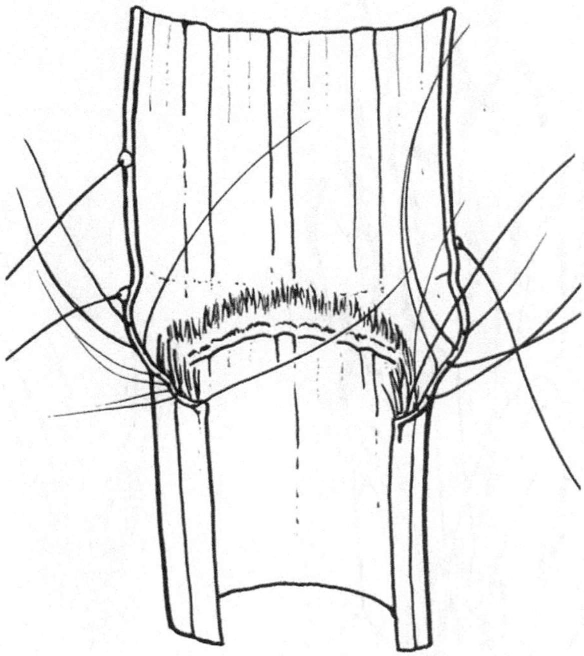

Fig. 9.204 Open sheath, collar, hairy ligule, and blade of Sixweeks Needle Grama (Copple and Pase 1967. U.S. Forest Service)

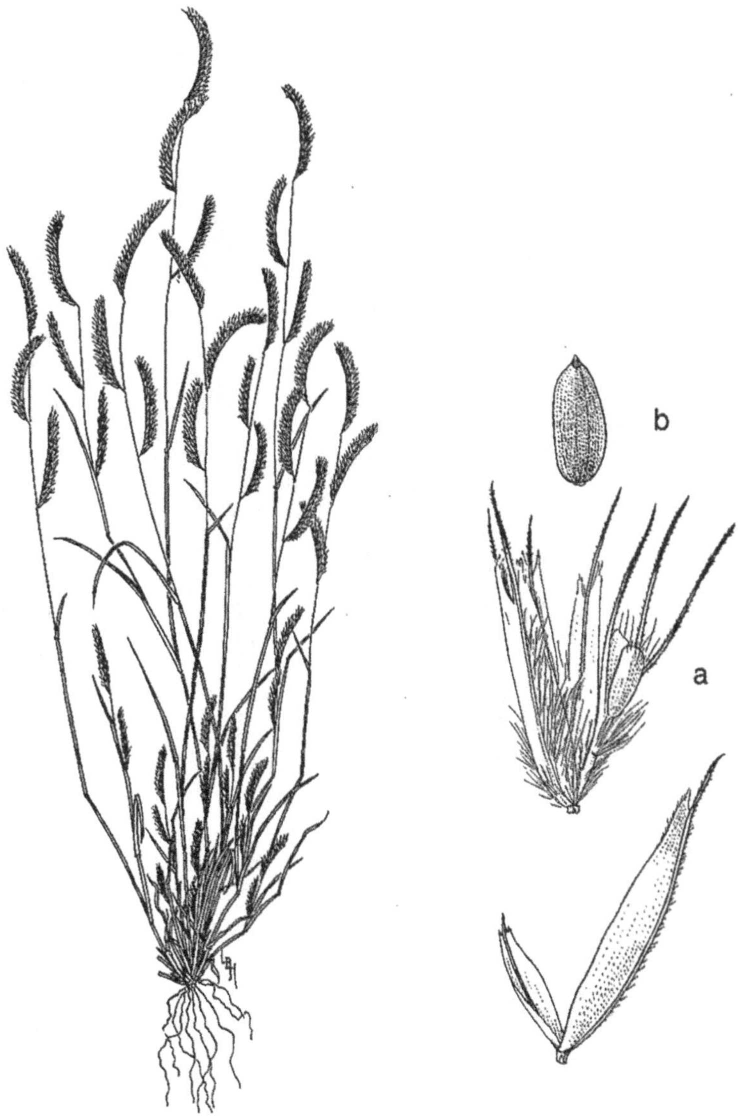

Fig. 9.205 Sixweeks Grama (*Bouteloua barbata*)—POACEAE. Annual 1–75 cm tall, prostrate or erect, some roots from lower nodes. **Leaves**: Hairless sheaths except for tufts of long hairs on both sides of collars; ligules 0.1–1 mm, ciliate; blades 5–10 mm long, 1–4 mm wide. **Flowers**: Panicles 7–25 mm with 2–10 branches 10–30 mm long with 20–50 spikelets. Each head ending in well-defined spikelet. **a**. Spikelet with fertile and sterile flowers with the separated glumes below. **b**. Grain (Fig. 9.206 from *An Illustrated Guide to Desert Weeds*, by Kittie F. Parker. © 1972 The Arizona Board of Regents. Reprinted by permission of the University of Arizona Press)

Sixweeks Grama (*Bouteloua barbata*) [A|DN]

For me, the *Boutelouas* are the grasses of the Southwest. Sod-forming perennials and ruderal annuals, they are a vital element of native ecosystems. Their leaves and seeds feed birds, small mammals, and ants. Alien introductions such as Red Brome will replace many of them, but some could survive. Three of them including Sixweeks Grama, its close relative, Rothrock Grama (*B. rothrockii*), and painful Sixweeks Needle Grama (*B. aristidoides*), described above, tolerate the heat of the Sonoran Desert and the freezes of the middle elevations around CWF. All three might persist while many other grasses decline or disappear altogether (Fig. 9.206).

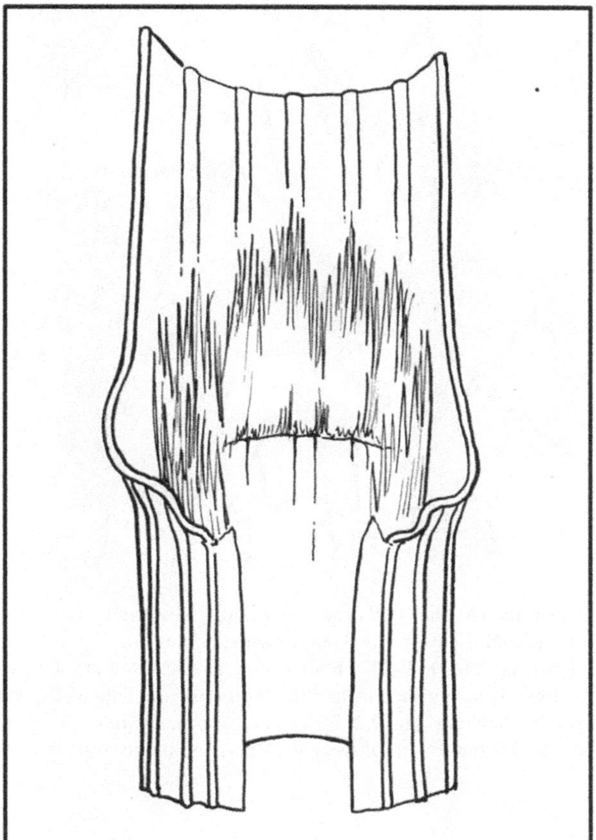

Fig. 9.206 Sixweeks Grama sheath, collar, ligule, blade (drawing by Copple and Pase 1967. U.S. Forest Service)

Fig. 9.207 Rescue Brome (*Bromus catharticus*)—POACEAE. Annual or biennial 6–36 in. high reproducing only from seed. **Leaves**: Flat 3–8 mm wide; ligules long, up to 5 mm, unfringed, but with some teeth. **Flowers**: Panicle 9–28 cm long with spikelets 2–4 cm long with 5–12 florets. **Grain**: about 1 cm long. Distinguished from other bromes by the flattened spikelets and short or absent awns. **a**. Spikelet. **b**. Grain (Fig. 9.207 from *An Illustrated Guide to Desert Weeds*, by Kittie F. Parker. © 1972 The Arizona Board of Regents. Reprinted by permission of the University of Arizona Press)

Rescue Brome (*Bromus catharticus*) [A-P|E]

This South American native is now present around the world. Also known as rescuegrass and prairie grass, Rescue Brome does well in dry locations. Though it is not tolerant of fire, its fine fuel production qualifies it as a member of the fuelweed group responsible for decreasing fire-return intervals and wiping out native species in arid Western North America. Though not as prolific as cheatgrass, it acts as a supporting character in the war on biodiversity loosed by humanity. It invades lawns and gardens, but regular mowing or pulling will give you control (Fig. 9.208).

Fig. 9.208 Rescue Brome spikelets (photo © Forest and Kim Starr)

Fig. 9.209 Red Brome (*Bromus madritensis* ssp. *rubens*)—POACEAE. Annual grass 10–40 cm tall. **Leaves**: Sheaths hairy, ligule 1–4 mm, hairy, pointed with smooth edges, blades 15 cm long and 1–5 mm wide, flat, hairy top and bottom. **Flowers**: Panicles 2–10 cm long and 2–5 cm wide, reddish-brown; spikelets 18–25 mm with 4–8 florets, lower glumes hairy 5–8 mm, upper glumes hairy 8–12 mm, 3–5 veins; lemmas 10–15 mm hairy; awns 8–20 mm; awns 2 cm long, curving as plant dries (Fig. 9.209 photo U.S. National Park Service)

Red Brome (*Bromus madritensis* ssp. *rubens*) [A|I]

This European annual is abundant west of the Rocky Mountains. It flourishes in the Arizona Uplands of the Sonoran Desert and the semiarid uplands of Central Arizona and Southern California. Like its cousin, Cheatgrass, Red Brome is a winter annual that completes its lifecycle by late spring, dries out, and stands waiting for a spark to burn in a hot flash across the land. It is not the worst fuel plant in the region, Buffelgrass (*Cenchrus ciliaris*) invading the desert south of CWF claims that honor, but when combined with other alien weeds, Red Brome is a significant contributor to destructive wildfires (Fig. 9.210).

Fig. 9.210 Maturing stand of Red Brome (photo © GR)

Fig. 9.211 Cheatgrass or Downy Brome (*Bromus tectorum*)—POACEAE. Erect stems. **Leaves**: Hairy with 2–3 veins on each side of the midrib; ligule small 1–2 mm, hairless, membranous, obtuse-lacerate. **Seeds:** Heads drooping with maturity with 1–2 cm long awns on the lemma (Fig. 9.211 drawing Hitchcock and Chase 1950, public domain)

Cheatgrass (*Bromus tectorum*) [A|IP]

Shoes become bushy during a summer stroll through a field of Cheatgrass. Designed for mammal dispersal, the plant's rough seeds burrow between shoelaces, around shoe tongues, and into socks where they poke and scratch. However, they do not generate the emergency reaction of the larger harder seeds of Ripgut Brome described next.

The ecological pain caused by Cheatgrass is much greater than the physical pain it administers to animals. This Eurasian annual germinates in winter or early spring and completes its life cycle by early summer. Through the rest of the summer and early fall it stands crispy dry ready to burn. Ignitions come from lightning, campfires, trains, careless smokers, and hot car mufflers. Once ignited, cheatgrass fires devour the land with intense heat that kills soil organisms and ignites herbs, shrubs, and trees. A botanical irony, many cheatgrass seeds survive the fire. The seeds germinate the next spring and produce new plants and more seeds wind and mammals spread across surrounding areas. After a few years and a few fires, almost nothing but Cheatgrass remains. This fire-resistant little grass has depleted biodiversity and productivity of millions of acres of native vegetation in the Western United States.

Enterprising American ranchers further deplete diversity by trucking in cattle to graze cheatgrass ranges in spring and then trucking the cows out as soon as the grass dries in early summer. Wildlife and feral horses struggling to survive on the impoverished land find and consume every plant on every fireproof rocky mound leaving few native plants for future recovery (Fig. 9.212).

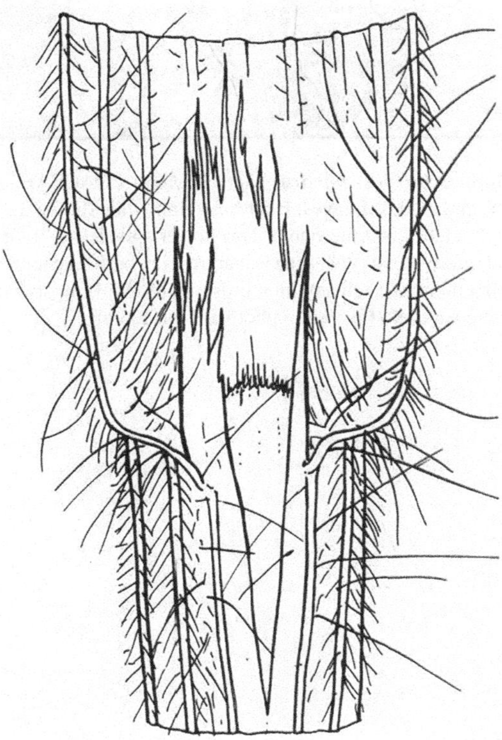

Fig. 9.212 Cheatgrass sheath, collar, ligule, and blade (drawing by Copple and Pase 1967. U.S. Forest Service)

Fig. 9.213 Ripgut Brome (*Bromus diandrus* ssp. *rigidus*)—POACEAE. Annual with smooth stems and hairy sheaths and blades. **Leaves**: Blades 3–6 mm wide; ligule 2–5 mm, tattily toothed. **Flowers** 10–20 cm on branches bearing one or two flower spikelets 3–4 cm long. Awns 3–6 cm long. **Fruit**: Ellipsoidal, grooved caryopsis hairy at apex. Tiny barb-like hairs that point backwards cover the seeds, allowing them to catch anything brushing by. Movement works the sharp-pointed seeds into fur, skin, and even eyes (Fig. 9.213 photo © Matt Lavin)

Ripgut Brome (*Bromus diandrus* ssp. *rigidus*) [A|IP]

This is the most painful grass I have encountered. Unfortunately, it has claimed the area beneath the Cottonwood trees along the river. It germinates in early spring and covers the ground beneath the trees with lush waves of green that give no hint of the agony to come. The seed heads mature in May. Thereafter, the river is almost safe from marauding humans. One person who I repeatedly asked to use the streets for his walks and leave the river woods for wildlife was self-important enough to grumble even with his snow gaiters on, the grass was too painful for pleasant walks anyway.

Ripgut in yards and gardens is easy to control by mowing and pulling. Periodic floods deposit rocks and debris over the river pasture and make mowing almost impossible. I wear snow gaiters when I want to spend time in that pasture during summer (Fig. 9.214).

I do not know for sure Ripgut seeds harm wildlife, but I hope most furry creatures have learned how to avoid the seeds or how to remove them.

Fig. 9.214 Ripgut Brome and Cheatgrass (*Bromus tectorum*). Revegetation of disturbed natural areas is labor intensive and often fails because of the relentless arrival of invasive weeds (photo © Matt Lavin)

Fig. 9.215 Field Sandbur (*Cenchrus spinifex*)—POACEAE. Spreading annual forms mats 4–36 in. wide with ascending leaves and flower spikes. **Leaves**: Two to five inches long; ligule ciliate, membranous base very short, collar margins villous (long soft straight hair). **Flowers**: Spikes 4–10 cm, often partly enclosed by upper leaf sheath and holding 3–15 loosely arranged burs but some with 20–30 tightly packed burs. **Seeds**: Spiny, hairy, yellowish burs, about 13 mm long, contain two seeds. Up to 500 burs per plant. **a**. Bur with spikelets **b**. Top and bottom views of grain. **c**. Bur from the similar Southern Sandbur (*Cenchrus echinatus*) that may not reach CWF until the climate warms more (Fig. 9.215 from *An Illustrated Guide to Desert Weeds*, by Kittie F. Parker. © 1972 The Arizona Board of Regents. Reprinted by permission of the University of Arizona Press)

Field Sandbur (*Cenchrus spinifex*) [A|P]

This grass forms seeds with sharp spines that can exceed one-half inch. Not as terrible as Bullhead perhaps, but right up there. The first time I encountered Field Sandbur, I was wearing rubber wader boots while fixing an aerator line in a pond. After walking up the bank from the pond, I noticed seeds stuck to one boot. I brushed them off with a bare hand—yikes!—experience is a great teacher. The red of the young burs in the photo below is an appropriate warning (Fig. 9.216).

Fig. 9.216 Field Sandbur fruit (photo © Macleay Grass Man)

Fig. 9.217 Feather Fingergrass (*Chloris virgata*)—POACEAE. Slender spreading annual 6–40 in. high. **Leaves**: Blades 2–8 cm long, with upper leaf bases wrapped around immature flower heads. **Ligule**: Membranous, 1 mm, truncate-ciliate. **Flowers**: Two to ten soft feathery spikes 2–8 cm long arising from stem tips. Each spikelet has a tuft of long soft hair and very slender bristles 6–10 mm long. Spikelets crowded in two rows along one side of the spike stem. **a**. Spikelet, glume detail below (Fig. 9.217 from *An Illustrated Guide to Desert Weeds*, by Kittie F. Parker. © 1972 The Arizona Board of Regents. Reprinted by permission of the University of Arizona Press)

Feather Fingergrass (*Chloris virgata*) [a|N]

This native annual grass has spread worldwide to regions with hot summers. People carried the plant to Australia, but botanists believe it must have dispersed across the oceans to other regions before people built seafaring ships. Such long-range dispersal is rare, but not impossible (Ridley 1930). One force for dispersal of small seeds like Feather Fingergrass is wind. Unusual winds can carry seeds aloft for great distances. Likewise, naturalists have reported rafts of matted water plants far from land carrying grasses, shrubs, and even small trees. Through the millions of years since the grasses appeared, such long-distance transport must have occurred thousands of times.

Bermudagrass (next species) resembles Feather Fingergrass, but Bermudagrass is smaller and perennial with many aboveground stolens and belowground rhizomes.

Feather Fingergrass is an excellent colonizer. Like magic, it spreads into disturbed areas including lawn margins, gardens, cultivated fields, roadsides, stream channels, and overgrazed livestock ranges (Fig. 9.218).

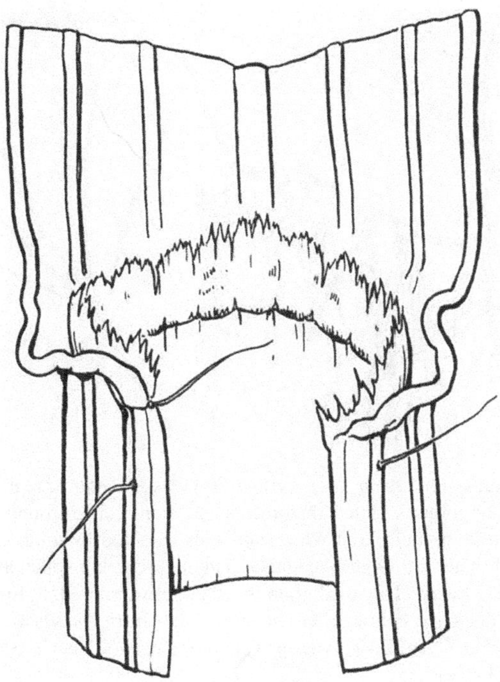

Fig. 9.218 Feather Fingergrass sheath, collar, ligule, blade (drawing by Copple and Pase 1967. U.S. Forest Service)

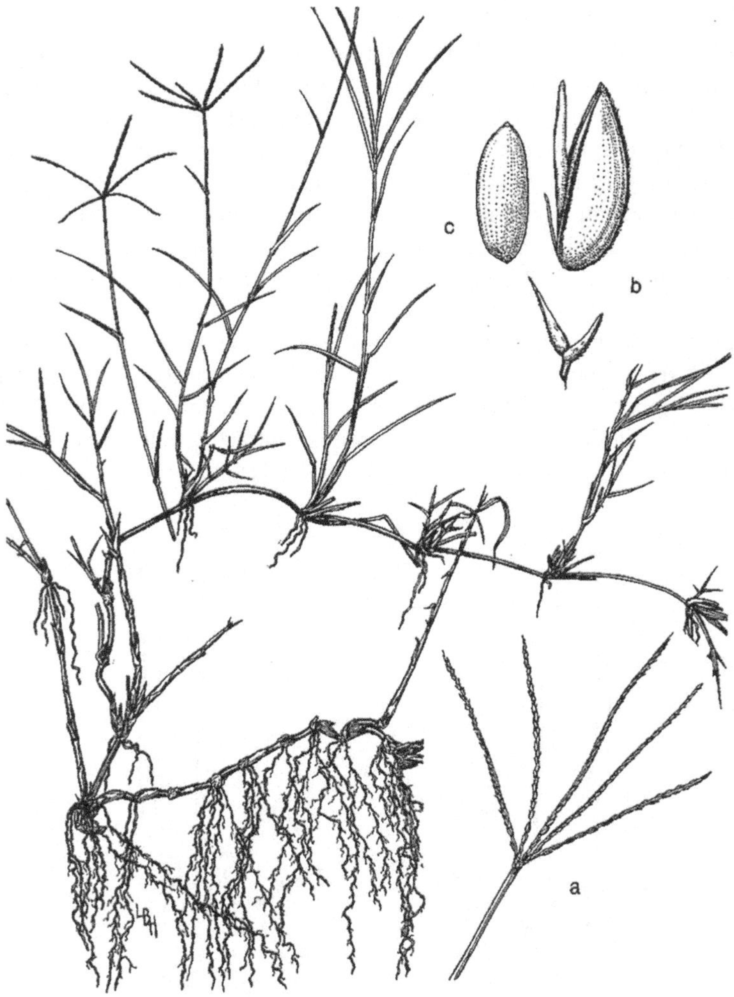

Fig. 9.219 Bermudagrass (*Cynodon dactylon*)—POACEAE. Plants spread by runners atop the ground that root at their joints. Pointed rhizomes spear along underground. **Leaves**: 1–4 inches long. The ligule 1 mm long, a circle of white hair with scattered long hairs. **Flowers**: On stalks form 4–18 inches high with 3–5 fingerlike spikes. Tiny overlapping spikelets appear in two rows on one side of the spike. **Seeds**: Tiny oval grain is red or straw colored. **a**. Enlarged inflorescence. **b**. Seed with glumes detached below. **c**. Grain (Fig. 9.219 from *An Illustrated Guide to Desert Weeds*, by Kittie F. Parker. © 1972 The Arizona Board of Regents. Reprinted by permission of the University of Arizona Press)

Bermudagrass (*Cynodon dactylon*) [P|DT]

Bermudagrass forms thick turf on Caribbean islands, but is a newcomer there as it is in lawns throughout Arizona. Botanists believe the grass originated in China.

Bermudagrass is so tough some people call it Devil Grass. It is a problem if it gets established where you do not want it. Its roots go deep. They will send up new shoots long after you have removed all the aboveground parts. And you cannot use the aboveground parts, the stolens, for mulch; with the next rain or irrigation, the stolens will form new roots and resume growth.

Bermudagrass makes a fine ground cover that needs almost no care. It does well under eves where rainwater concentrates, and it will grow in other places where it catches a little overshoot from lawn sprinklers. Butterflies and Damselflies find the grass interesting, and Crocuses do well in Bermudagrass lawns. The bulbs flower before the grass shakes its winter depression and they mature by time to mow. Gophers and Javelina love the bulbs from which these flowers grow, but they always leave some for next spring (Fig. 9.220).

Some people have a mild itchy reaction to Bermudagrass, but soap and water cures the itch. SEINet says the leaves are rich in vitamin C and make a fine herbal tea. I have not tried the tea, but I know dogs and cats eat the leaves.

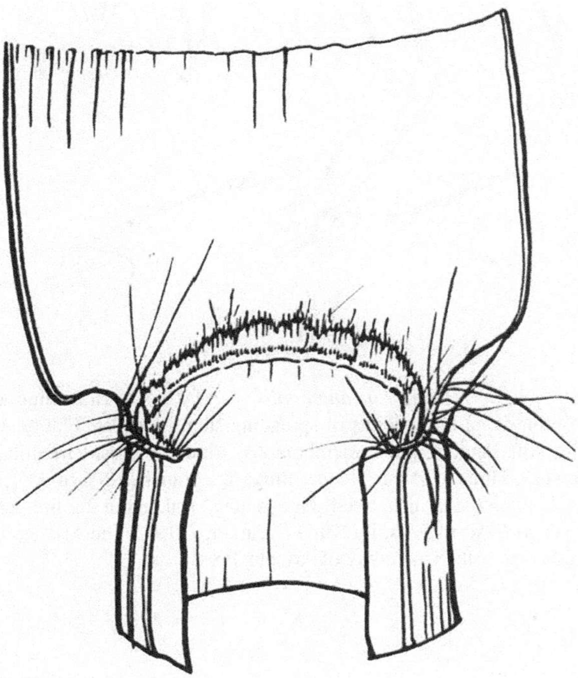

Fig. 9.220 Bermudagrass sheath, collar, ligule, blade (drawing by Copple and Pase 1967. U.S. Forest Service)

Fig. 9.221 Large Crabgrass (*Digitaria sanguinalis*)—POACEAE. Branching annual reproduces by seed and by rooting from lower joints of spreading stems. **Leaves**: Flat blades 6–8 mm wide. Sheaths with long stiff hairs. Ligules membranous, truncate, erose (irregularly toothed) 0.5–2.5 mm long. **Flowers**: Three to eleven slender fingerlike branches 2–6 in. long have small spikelets 3–6 mm long that hug the branch. **Seed**: Pale yellow, oval, about 2 mm long (Fig. 9.221 from *An Illustrated Guide to Desert Weeds*, by Kittie F. Parker. © 1972 The Arizona Board of Regents. Reprinted by permission of the University of Arizona Press)

Large Crabgrass (*Digitaria sanguinalis*) [A|T]

Large Crabgrass requires moist soil. It colonizes lawns, banks of the Agua Fria River, and low spots where water collects. Plants form a basal rosette of leaves low enough to escape mower blades. If you need a perfect lawn, pull the plants or dig them up (Fig. 9.222).

Fig. 9.222 Large Crabgrass (photo © Rasbak)

Fig. 9.223 Barnyard Grass (*Echinochloa crus-galli*)—POACEAE. This strong summer annual matures anywhere from 6 in. to 72 in. tall and often spreads from its base to form large clumps. The leaves are 4–20 in. long and 6–20 mm wide; *ligules absent*, ligule region sometime hairy. Flowering occurs along the top 3–10 in. of the stem. Flowering branches are 1–2 in. long with short green or purple spikelets 3 mm long with bristles up to 4 cm long. A single plant can produce 40,000 of the pale yellow shiny seeds. **a**. Spikelet spread open. **b**. Two spikelets with short and long awns. **c** Grain (Fig. 9.223 from *An Illustrated Guide to Desert Weeds*, by Kittie F. Parker. © 1972 The Arizona Board of Regents. Reprinted by permission of the University of Arizona Press)

Barnyard Grass (*Echinochloa crus-galli*) [A|D]

Farmers detest this European grass for its ready invasion and persistence in most crops. A single plant can produce 40,000 seeds. In the arid climates, the plant survives along moist stream banks and in moist spots beside roads, lawns, and gardens. Here, the plant is easy to control by mowing, pulling, or withholding water. However, if you allow a few clumps to grow, you will get to enjoy watching the acrobatics of Lesser Goldfinches that relish Barnyard Grass seeds. The small birds hang (often upside down) on drooping seed tops removing mature seeds until all are gone. The birds must drop seeds, however, as new plants appear in the same spots every year (Fig. 9.224).

Fig. 9.224 Barnyard Grass with Lesser Goldfinch showing his tail feathers (photo © GR)

Fig. 9.225 Junglerice (*Echinochloa colona*)—POACEAE. Annual 9–36 in. tall reproducing from seeds and spreading stems. **Leaves**: Narrow, 2–10 mm wide, 8–22 cm long with *no ligules*. First leaves often form a dense rosette. Most leaves have distinct purple bands. **Flowers**: In panicles 2–12 cm long with 2–10 branches, spikelets hairy, 2–3 mm wide, and often purple. **A**. grain (Fig. 9.225 from *An Illustrated Guide to Desert Weeds*, by Kittie F. Parker. © 1972 The Arizona Board of Regents. Reprinted by permission of the University of Arizona Press)

Junglerice (*Echinochloa colona*) [A|E]

No one knows for sure where this plant originated, but the best guess is it evolved and spread from Old World tropics. It invades disturbed moist sites worldwide. The purple bands on the leaves are unusual. Their ecological purpose eludes me, but they do simplify identification. An invader of rice fields, Junglerice is itself cultivated for food in India and Egypt (Fig. 9.226).

Fig. 9.226 Junglerice (Correll and Correll 1972, Public Domain). The artist did not illustrate the purple bands on the leaves. Perhaps they fade as collected plants dry. **A.** hairy leaf blade connection to stalk; **b.** fruit; **c.** seed; **d.** grain; **e.** roots

Fig. 9.227 Stinkgrass (*Eragrostis cilianensis*)—POACEAE. Eurasian annual, branches from the base, culm 4–24 in. long, prostrate then ascending. **Stems and leaves:** There is often a row of glandular pits below the joints and others scattered on leaves and flower stalks. The pits may produce an odor that repels livestock, but many people can't detect the odor. Ligule is a fringe of short hairs (ciliate) less than 1 mm long. **Flowers**: Spreading 2–10 in. long with 12–40 crowded on spikelets 6–20 mm long, pointed at both ends. **Seed**: Brown oval grains 1–2 mm long pointed at both ends. **a**. Stem with ring of (stink?) glands below the joint. **b**. Spikelet. **c** Grain (Fig. 9.227 from *An Illustrated Guide to Desert Weeds*, by Kittie F. Parker. © 1972 The Arizona Board of Regents. Reprinted by permission of the University of Arizona Press)

Stinkgrass (*Eragrostis cilianensis*) [A|I]

The malodorous defense. Odor produced by glandular pits on this European weed reminds some people of cockroaches. The similarity is not clear to me. Stinkgrass has not been abundant at CWF, but it made a major expansion after 2018. If the local gopher/mouse crowd will eat it, it gets to stay (Fig. 9.228).

Fig. 9.228 Stinkgrass at CWF (photo © GR)

Fig. 9.229 Lehmann's Lovegrass (*Eragrostis lehmanniana*)—POACEAE. Perennial *to 1 m tall*. **Leaves**: Ligules 0.3–.5 mm ciliate, blades *2–12 cm long*, 1–4 mm wide. **Flowers**: Open panicles 7–18 cm long and 2–8 cm wide with branches 1–8 cm long and spikelets 5–14 mm long and about 1 mm wide with 4–14 florets. **Seeds**: Caryopses about 1 mm long, hairless, light brown, seed dark brown (Fig. 9.229 photo © Max Licher)

Lehmann's Lovegrass (*Eragrostis lehmanniana*) [P|I]

Most weeds disperse in wind, water, and animal fur. However, people transport and plant some of them to improve on the services provided by native plants. Range managers introduced two Lovegrasses (Lehmann and Weeping) from South Africa to stabilize soils in overgrazed rangelands. Land managers also planted the grasses to stabilize soil in roadcuts, transmission corridors, and elsewhere. During the 1980s ecologists realized the grasses were invading adjacent native vegetation. Within a few years, public agencies stopped planting Lovegrasses, but the plants were already out of control. They cover over 20,000 acres of Arizona's arid grassland and they are still spreading.

Lehmann's Lovegrass is unlikely to invade the Desert Grassland or Interior Chaparral near CWF. It appears to need more precipitation than we receive. One patch in the edge of an irrigated pasture at CWF spreads to about 500 ft. square by the time I stopped irrigating the pastures in 2009. Without irrigation and with competition from Kochia, the plants faded away until now, 8 years after I turned off the water, there are only a few tufts left under the Cottonwood Trees where Kochia does not grow (Fig. 9.230).

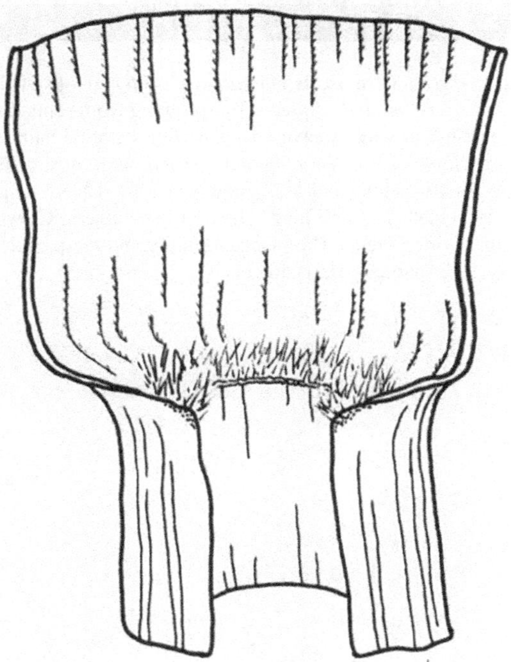

Fig. 9.230 Lehmann's Lovegrass sheath, ligule, blade (drawing by Copple and Pase 1967. U.S. Forest Service)

Fig. 9.231 Weeping Lovegrass (*Eragrostis curvula* var. *conferta*)—POACEAE. *Upright* perennial, *45–190 cm tall*; *tufts form mounds*. Spreads by sprouting from roots that penetrate over 4 m deep and spread more than 3 m wide. **Leaves**: Sheath with scattered hairs, ligules ciliate, membrane about 1 mm long. *Blades 12–60 cm long* and 1–3 mm wide, scattered hairs to 7 mm long. **Inflorescence**: Panicle 16–40 cm long and 4–24 cm wide with 4–10 mm long spikelets supporting 3–10 florets. Glumes lanceolate 1–3 mm long. Three brown anthers. **Caryopses**: 1–2 mm long, smooth, light brown, base often green. The plants are taller and the panicle is denser than other Eragrostis species (Fig. 9.231photo © Harry Rose)

Weeping Lovegrass (*Eragrostis curvula* var. *conferta*) [P|I]

Weeping Lovegrass is taller and has larger leaves than Lehmann's Lovegrass. Otherwise, the two are similar in form and behavior (Fig. 9.232).

Fig. 9.232 Weeping Lovegrass (photo © Harry Rose)

Fig. 9.233 Wild Barley (*Hordeum murinum* subsp. *leporinum*)—POACEAE. Annual, reproduces by seed, prostrate, with branches 6–20 in. long. **Leaves**: Flat, 4–10 cm long. Ligules short, membranous. **Flowers**: Erect spike 2–3 in. long, upper portion enclosed in leaf sheath. **Fruit**: Stiff bristles 2–4 cm long; seed about 6 mm long and hairy on top. **a**. Spikelets. **b**. Floret with long bristle (Fig. 9.233 from *An Illustrated Guide to Desert Weeds*, by Kittie F. Parker. © 1972 The Arizona Board of Regents. Reprinted by permission of the University of Arizona Press)

Foxtail Barley (*Hordeum jubatum*). Similar to Wild Barley, but perennial reproducing by seed and by sprouting from rhizomes. Flowering spikes larger and nodding. The seeds penetrate animal skin and ears, often found inside dog feet having entered through the skin between toes.

Wild Barley (*Hordeum murinum* subsp. *leporinum*) [A-P|PT]

Around CWF, Wild Barley and Foxtail Barley grow along the Agua Fria River, and on the margins of lawns and gardens. People plant Foxtail Barley for the beauty of its seed heads, but that is not a good idea if you have pets. The bristled seeds will burrow into fur and they will inch into nostrils, ears, and some will pierce the skin. When they break the skin, bacterial infections may develop (Fig. 9.234).

Both weeds are difficult to control. Wild Barley is difficult because of its heavy seed production and rapid spread. Foxtail Barley is difficult because of its tough spreading roots. In both cases, repeated mowing will do the trick though at first it may seem hopeless. Keep going and you will prevail.

Fig. 9.234 Wild Barley (photo © Rasbak)

Fig. 9.235 Witchgrass (*Panicum capillare*)—POACEAE. Hairy annual reproducing only by seed, 4–36 in. high. **Leaves**: Spreading, 3–18 mm wide, 5–40 cm long, covered by long hairs; ligule ciliate with cilia 0.5–1.5 mm long. **Flowers**: Panicle, 13–50 cm long, branching with one 2–3 mm long spikelet at tips of branches; panicle often breaks free and blows like a tumbleweed. **Seeds**: Shiny yellow or gray about 2 mm long. **a**. Spikelet. **b**. Seeds (Fig. 9.235 from *An Illustrated Guide to Desert Weeds*, by Kittie F. Parker. © 1972 The Arizona Board of Regents. Reprinted by permission of the University of Arizona Press)

Witchgrass (*Panicum capillare*) [A|N]

Witchgrass' thin spikes create a misty appearance from a distance. It grows in all the likely places along roads, ditches, and the margins of fields and gardens. One of the most graceful grasses, gardeners often plant Witchgrass in borders and beds. However, the panicles break free and tumble along spreading seeds. If unwelcome seedlings invade your garden, mulching, hoeing, and pulling will bring them under control.

The small seeds appeal to little songbirds, mice, and voles. The leaves appeal to everyone (Fig. 9.236).

Fig. 9.236 Witchgrass ripening panicle (photo © Jim Pisarowicz, U.S. National Park Service)

Fig. 9.237 Carolina Canarygrass (*Phalaris caroliniana*)—POACEAE. Bluish green annual to 4 ft. reproducing only by seed. **Leaves**: Flat or folded 2–15 mm wide. Ligules 1–7 mm flat to somewhat pointed with shredded edge. **Flowers**: One oblong panicle 1–8 cm long topping each stem. Thicker and stubbier than similar Littleseed Canarygrass. Spikelets crowded and overlapping. Two broad outer bracts (glumes) 3–6 mm long, sharp, flattened, and folded with upper part of fold forming a papery wing with dark green stripes on each side of the base. **Seed**: **a**. Spikelet with two glumes and seed. **b**. Shiny brown, hairy, 3–5 mm long. **c**. Flowering stem (Fig. 9.237 from *An Illustrated Guide to Desert Weeds*, by Kittie F. Parker. © 1972 The Arizona Board of Regents. Reprinted by permission of the University of Arizona Press)

Carolina Canarygrass (*Phalaris caroliniana*) [A|N]

This annual native grass grows on wet stream banks and other moist disturbed areas. It might show up in a lawn, but mowing will blend it with other grasses and eventually eliminate it (Fig. 9.238).

Fig. 9.238 Carolina Canarygrass, glumes, and floret (drawing in Hitchcock and Chase 1950, USDA)

Fig. 9.239 Annual Bluegrass (*Poa annua*)—POACEAE. This grass is never more than 1 ft. tall. **Leaves**: Soft green leaves have the closed boat-shaped tips of other Poas (bluegrasses). Leaf sheaths enclose the stems for about one-third their length; ligules 1–3 mm long, hairless, obtuse to flat. **Flower**: The inflorescence is a panicle 2–8 cm long with 2–6 florets on short spikelets. **a**. Spikelet. **b**. Grain (Fig. 9.239 from *An Illustrated Guide to Desert Weeds*, by Kittie F. Parker. © 1972 The Arizona Board of Regents. Reprinted by permission of the University of Arizona Press)

Annual Bluegrass (*Poa annua*) [A|I]

Annual Bluegrass has traveled with people from its Eurasian home to every continent. In America, Sereno Watson found it in the Great Basin Desert of Utah and Nevada during the 1872 exploration of the fortieth parallel. He "collected [it] on stream banks in cañons near roads traveled only by teams of woodmen" (Watson 1871, p. xli).

Exceptional summer monsoon and winter rains kept the ground muddy for months during my first year at CWF. Shoes had to remain outside the door at home and work. Unfamiliar weeds, some with irritating or painful seeds, filled the fields and bare areas around the house and barn. One weed, Annual Bluegrass, covered patches along the driveway. Soft and green that winter and spring, the grass formed a dry mat during the summer, providing a yearlong mud-free walkway.

What a great plant. I identified it using Parker's book (Parker 1972) and headed for town. I asked about seeds at the local building-supply where lawn-grass seeds sold. After a reflective moment, the store employee fulfilled his customer-service obligation without detectable sarcasm and stated, "It is a weed." No seed was available.

I found no Annual Bluegrass seed. After 20 years tending this place, I have covered the mud with gravel and perennial plants, but patches of the little grass are still present beside the driveway and always trying to creep into the lawn. I could till and reseed it away, but sparrows and butterflies like it, so I will let it be (Fig. 9.240).

Fig. 9.240 Annual Bluegrass (photo © James Lindsey)

Fig. 9.241 Rabbitfoot grass (*Polypogon monspeliensis*)—POACEAE. Yellow-green annual 4–24 in. tall with stems often bent and spreading at lower joints, reproduces only from seed. **Leaves**: Blades 2–10 mm wide, sheaths open or loose, smooth, hairless. Ligule acute, unfringed. **Flowers**: Spike of many tiny spikelets about 2 mm long. Two outer bracts in each spikelet end in a bristle 6–10 mm long. Lemma and palea thin, translucent. **Seed**: Plump, amber, caryopsis 1–2 mm long. **a.** Spikelet. **b**. Spikelet of Ditch Polypogon (*P. interruptus*)

Ditch Polypogon (*Polypogon interruptus*). -Very similar to rabbitfoot grass, but spikes spread more. Bristles are shorter, about 3 mm long. (Fig. 9.241 from *An Illustrated Guide to Desert Weeds*, by Kittie F. Parker. © 1972 The Arizona Board of Regents. Reprinted by permission of the University of Arizona Press).

Rabbitfoot Grass (*Polypogon monspeliensis*) [A|I]

The origin of this furry moisture-loving grass is not clear. It might have originated in Europe, but it became a global cosmopolitan so early in human history; it is hard to retrace its path. I first discovered it growing from a cracked rock in the Agua Fria River. Traveling thousands of miles over oceans and mountains to get to this tiny stream running through a desert valley is the nature of weeds, but marvelous nevertheless (Fig. 9.242).

Fig. 9.242 Rabbitsfoot grass on a fractured rock in the Agua Fria River (photo © GR)

Fig. 9.243 Japanese Bristlegrass (*Setaria faberi*)—POACEAE. Annual 50–200 cm tall. **Leaves**: Flat, alternate 15–30 cm long, 1–2 cm wide. Membranous ciliate ligules 2 mm long. **Flowers**: Bristly arching spike 6–20 cm long, crowded by spikelets 2–3 mm long. *Five to twenty bristles arise from the base of each spikelet.* **Seeds**: Oval, dark green to brown about 2 mm long, flattened on one side (Fig. 9.243 photo by Mohlenbrock 1989, USDA)

Japanese Bristlegrass (*Setaria faberi*) [A|EI]

The bristles surrounding the seeds are not hard and sharp, making the grass more attractive than threatening. The real problem is the plant's ability to invade natural vegetation, which brings us to the questions: *What and where is natural vegetation?* The short answer is *natural vegetation is wherever native plants dominate and persist.* Something to consider, however, is that a thousand-pound cow has stepped on almost every square foot of the land. In most locations, the cows have weakened native plants and broken biological soil crusts as thoroughly as if someone plowed and replanted the land. The ungrazed places I have seen have few invasive weeds like Japanese Bristlegrass. However, as long as we continue eating beef and wearing belts and boots of cow skin, the ungrazed places will shrink (Fig. 9.244).

Fig. 9.244 Japanese Bristlegrass (photo © GR). The arching spikes dry and continuing dropping seeds through winter months

Fig. 9.245 Johnsongrass (*Sorghum halepense*)—POACEAE. Perennial 3–10 ft. tall spreads by seed and underground rhizomes. **Leaves**: Bright green to 2 ft. long, 6–20 mm wide. Ligule ciliate, 2–6 mm long. **Flowers**: In open branching tops 6–24 in. long. Spikelets in pairs, in threes at tips of branches. One or two stalked flowers bear stamens, the other one is about 6 mm long, stalkless, thicker, fertile. **Seeds**: Dark brown to black grains 3 mm long, shiny. **a**. Three spikelets, two sterile and stalked, the third fertile and awned, at the tip of the stem. **b**. Grain with hull (Fig. 9.245 from *An Illustrated Guide to Desert Weeds*, by Kittie F. Parker. © 1972 The Arizona Board of Regents. Reprinted by permission of the University of Arizona Press)

Johnsongrass (*Sorghum halepense*) [P|I]

Johnsongrass a perennial introduction of Mediterranean origin. There are several small patches at CWF. Some of the patches receive a little overspray from lawn sprinklers and others are close to the river. Annual mowing and Javelina rooting for rhizomes easily control them. The plant first appeared by the river 12 years ago. It could be a problem now if I still irrigated any pastures, but that ended when my last rescue cow died in 2009 (Fig. 9.246).

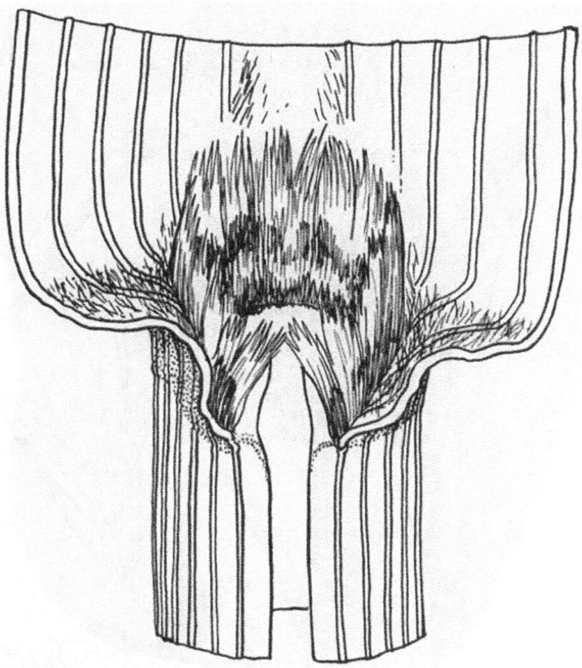

Fig. 9.246 Johnsongrass sheath, collar, ligule, blade (drawing by Copple and Pase 1967. U.S. Forest Service—secret Wookiee fans?)

Fig. 9.247 Giant Reed (*Arundo donax*)—POACEAE. Plants 2–5 m tall (sometimes taller) in clumps. Hollow stems 2–3 cm in diameter. **Leaves**: Evenly spaced, alternate, 30–60 cm long, 2–7 cm wide, with rough edges and wedgeshaped brown mark at base. Ligules an unfringed membrane <1 mm. **Flowers**: Feathery plumes 30–60 cm long and up to 30 cm wide. Spikelets 10–15 mm long with 3–6 flowers seedless or with infertile seeds. Reproduces from underground rhizomes (Fig. 9.247 drawing from Hitchcock and Chase 1950, USDA)

Giant Reed (*Arundo donax*) [P|I]

Giant Reed has not colonized the farm yet, but the plant is present elsewhere in the area. It is invasive in moist locations, but can be controlled in our arid climate by withholding water. However, it might become a problem along the River. According to contributors to the IUCN Global Invasive Species Database, this is one of the 100 world's worst weeds (International Union for the Conservation of Nature (IUCN), Invasive Species Specialist Group. 2019). Worth watching for.

Grazing animals can eat young plants, and some wildlife species like to rest and nest in the clumps formed by this big leafy grass. The leaves and stems contain high silica content making them quite durable. People have used them for thousands of years for building material, fishing poles, walking sticks, panpipes, and flutes and it is the principal source of reeds for wind instruments (Fig. 9.248).

Fig. 9.248 Giant Reed (photo © Forest and Kim Star)

Fig. 9.249 Crested Wheatgrass (*Agropyron cristatum*)—POACEAE. To 110 cm tall, often under 40 cm, spikes 10 cm with two rows of overlapping spikelets. **Leaves**: 5–20 cm long, 2–6 mm wide; ligules membranous, less than 2 mm. **Flowers**: Spikes to 10 cm long with 2 rows of overlapping 7–16 mm long spikelets each with 3–6 florets; anthers 3–5 mm. **Fruit**: Elliptical grooved caryopsis. **Insets** left to right: Flower, fruit, stigma, grain (Fig. 9.250 painting by Schmidt and Ibmayer 1801-1809)

Crested Wheatgrass (*Agropyron cristatum*) [P|I]

U.S. range managers imported this Siberian steppe grass to plant on overgrazed ranges stripped bare by cattle, fire, or construction (Dillman 1946). Crested Wheatgrass grows well on bare ground, protects the soil, and satisfies the need for a nutritious feed for livestock (Figs. 9.250 and 9.251).

Range managers soon discovered the grass they planted blocked the return of native plants for a low diversity landscape of reduced value for wildlife. Now the plant is the dominant on over ten million acres in the northern Great Plains and Canada. It surprised me to find it growing this far south. Control will not be a problem in our warm dry landscape.

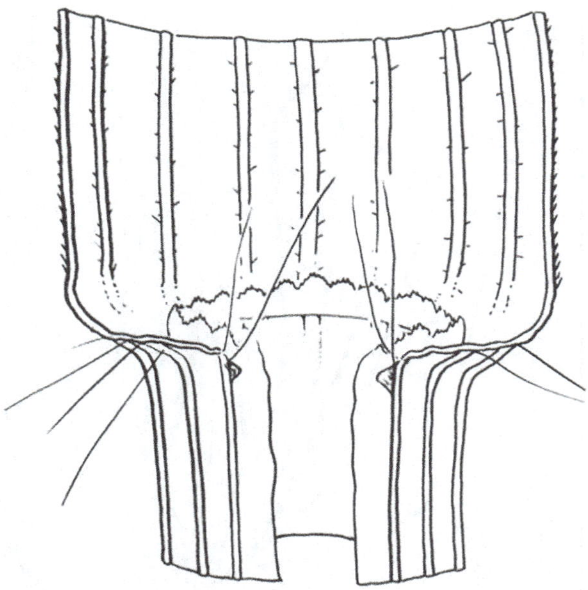

Fig. 9.250 Crested Wheatgrass sheath, ligule, blade (drawing in Copple and Pase. 1967. U.S. Forest Service)

Fig. 9.251 Crested Wheatgrass at CWF (© Max Lavin)

Fig. 9.252 Jointed Goatgrass (*Aegilops cylindrica*)—POACEAE. Annual to 1 m tall. **Leaves**: Open sheaths; ligule short, blades flat, slightly hairy, 3–14 cm long, and 2–5 mm wide. **Flowers**: Spikes 2–12 cm long, 3 mm wide, cylindrical, with 3–8 fertile spikelets 9–12 mm long. **Seeds**: Glumes of lower spikelets with 2–5 mm long awns, apical spikelet glumes with 3–6 cm awn. Lemmas of apical spikelets have one 4–8 cm long awn (Fig. 9.252 painting: artist unknown, Public Domain)

Jointed Goatgrass (*Aegilops cylindrica*) [AZ2]

An introduced Eurasian annual, Jointed Goatgrass' tight cylindrical seed head is distinctive. The plant is present in disturbed sites, especially along roads. In the Midwestern United States, it invades winter wheat and other crops. It has spread in some of the remnant grasslands of the Great Plains, but it has not invaded native desert grassland and shrublands in the Western United States.

Seeded joints of the grass often mix with wheat stems and the seeds blend with wheat seeds transported to new fields. Eradication efforts use crop rotations and seasonal applications of specialized herbicides, species-specific deleterious micro-bacteria, and burning, but nothing has been completely effective.

Though Jointed Goatgrass' production of food is about half that of winter wheat, the plant might become important for its greater resistance to drought (Fig. 9.253).

Fig. 9.253 Jointed Goatgrass (photo © Stefan.lefnaer)

Fig. 9.254 Quackgrass (*Elytrigia repens*)—POACEAE. Reproduces from seeds and rhizomes. Grows 1–5 ft. tall. **Leaves**: 6–30 cm long, 3–10 mm wide; ligules 0.3–1.5 mm. **Flowers**: Spikes 5–15 cm long and 5–15 mm wide with 10–27 mm spikelet per node. Spikelets fall intact (Fig. 9.254 painting: Normal Criddle in Clark and Fletcher 1909)

Quackgrass (*Elytrigia repens*) [P|IMTZ-2]

A Eurasian perennial scattered here and there around the farm, Quackgrass creeping rhizomes spread into lawns and into native vegetation. As with most grasses, the entire plant serves as food for wildlife species. Nevertheless, this alien grass is too destructive to ignore. Eradication is difficult, because the rhizomes weave among the roots of other plants and even a small fragment can develop into a new plant. Wikipedia has an extensive discussion of removal strategies. As with many other weeds, early detection is critical to control (Fig. 9.255).

Fig. 9.255 Blunt ligule 1 mm high (photo © Rasbak)

9.29 Knotweed Family: POLYGONACEAE (Fig. 9.256)

Fig. 9.256 Prostrate Knotweed (*Polygonum aviculare*)—POLYGONACEAE. Green or blue-green annual; sometimes whitish from powdery mildew. Reproduces by seeds. Branching stems form mats 4–48 in. across. **Leaves**: Small, alternate, pointed, 6–38 mm long, and up to 10 mm wide. Present to the ends of the stems. **Flowers**: **a.** White or green with pink tinge occur in small clusters in leaf axils on upper stems (Fig. 9.256). **Fruit**: **b**. Achenes three-sided dark brown to black, 1–4 mm long (Fig. 9.256 from *An Illustrated Guide to Desert Weeds*, by Kittie F. Parker. © 1972 The Arizona Board of Regents. Reprinted by permission of the University of Arizona Press)

Prostrate Knotweed (*Polygonum aviculare*) [A|N]

This tangle of thin stems lives beside the stepping-stones on my back walk. It cannot spread into the adjacent lawn because of periodic mowing. It does not impede traffic, and its tiny flowers and seeds feed a few arthropods and birds. The seeds remain viable for several years, so it should return every year (Fig. 9.257).

Some writers claim Prostrate Knotweed is native to both North America and Europe. Did the species spring up on both continents? The ancient Greek philosopher Aristotle might have accepted that possibility, but modern geographers know that species have to come from one place. While it is possible the tiny seeds crossed the ocean before humans came along to help, the SEINet distribution map for the species leads us to a much firmer conclusion. Looking at the SEINet map and applying a little spatial reasoning (Abler et al. 1971; Rappaport 1982), I say the plant is a North American native. Look at the map and think about seeds dispersing from a point of origin. What do you think?

Fig. 9.257 Prostrate Knotweed (photo © Stefan Lefnaer)

Fig. 9.258 Erect Knotweed (Wireweed) (*Polygonum erectum*)—POLYGONACEAE. A small lithe plant; grows 10–75 cm tall. **Leaves**: Green or yellowish, 3–6 cm long, 1–3 cm wide. Stem sheaths 7–12 mm long. **Flowers**: One to five tiny closed flowers in axils of most leaves. **Seed**: Achenes enclosed in perianth (Fig. 9.258 photo © GR)

Erect Knotweed (Wireweed) (*Polygonum erectum*) [A]

This Old World native is most common in Eastern United States and Canada. It colonizes disturbed areas but is not a significant invader of the livestock ranges of the west. It has been here for a long time; Serano Watson found it in Western Nevada in 1870 (Watson 1871).

The Knotweeds are easy to overlook. They are small, spindly, and have sparse foliage. This one grows beside my front walk. Eradication would be a simple matter of pulling the plant. However, it is an annual so I will leave it unless it spreads into the lawn. It is not in the way, and some members of the KNOTWEED FAMILY are favorites of butterflies and moths (Fig. 9.259).

Fig. 9.259 Erect Knotweed. Tiny flowers are present in the leaf axils (photo © GR)

Fig. 9.260 Red Sorrel (*Rumex acetosella*)—POLYGONACEAE. Perennial 6–24 in. tall reproduces from seed and rhizomes. **Leaves**: Alternate, arrowshaped 1–3 in. long with basal lobes, more numerous near base of plant. **Flowers**: Male flowers yellow, female reddish, on separate plants. **Seeds**: Achenes dark brown, 1–2 mm long. **a.** Male plant with staminate flowers. **b**. Branch from female plant in fruit. **c.** Achene with enclosed seed (Fig. 9.260 from *An Illustrated Guide to Desert Weeds*, by Kittie F. Parker. © 1972 The Arizona Board of Regents. Reprinted by permission of the University of Arizona Press)

Red Sorrel (*Rumex acetosella*) [P|EM]

Red Sorrel (Sheep Sour) from the southwestern Mediterranean region grows best in damp acidic soils. In CWF it grows on lawn margins and along the Agua Fria River. In our arid environment, the plant is easy to control. It does not spread across the land, and it dies after a few seasons of pulling or mowing. The weed scientist's call to duty, EDRR (early detection, rapid response) is certainly applicable; plants pulled before rhizomes spread may not sprout again (Fig. 9.261).

Fig. 9.261 Red Sorrell (photo © AnRo0002)

Fig. 9.262 Canaigre (Wild Rhubarb) (*Rumex hymenosepalus*)—POLYGONACEAE. Stout fleshy stems 1–3 ft. high, reproduces by seed and from rhizomes. Forms cluster of 2–12 tubers. **Leaves**: Margins smooth and wavy not as rough as Curly Dock (next). The leaves are hairless and have thin sheaths surrounding the base of the stalk. Lower leaves 8–24 in. long and 2–4 in. wide with straight edges. **Flowers**: Pale green in a dense leafless cluster 6–12 in. long. **a**. Fruiting calyx enlarged (Fig. 9.262 from *An Illustrated Guide to Desert Weeds*, by Kittie F. Parker. © 1972 The Arizona Board of Regents. Reprinted by permission of the University of Arizona Press)

Canaigre (Wild Rhubarb) (*Rumex hymenosepalus*) [P|EMN]

Canaigre is a native weed that grows in moist soils in disturbed sites. It grows beside roads, along the Agua Fria River, and in irrigated yards, gardens, and fields (Fig. 9.263).

Besides protecting soil and feeding wildlife, Canaigre has value for people. The roots contain tannin and produce a medium brown die. Aboveground plant parts are edible. Young leaves add tang to salads, and cooked older leaf stalks resemble rhubarb. I like the leaves, but not everyone does. "The leaves can be used as a substitute for rhubarb, although frankly it is pretty grim stuff" (Moore 2003).

Fig. 9.263 Canaigre © Stan Shebs

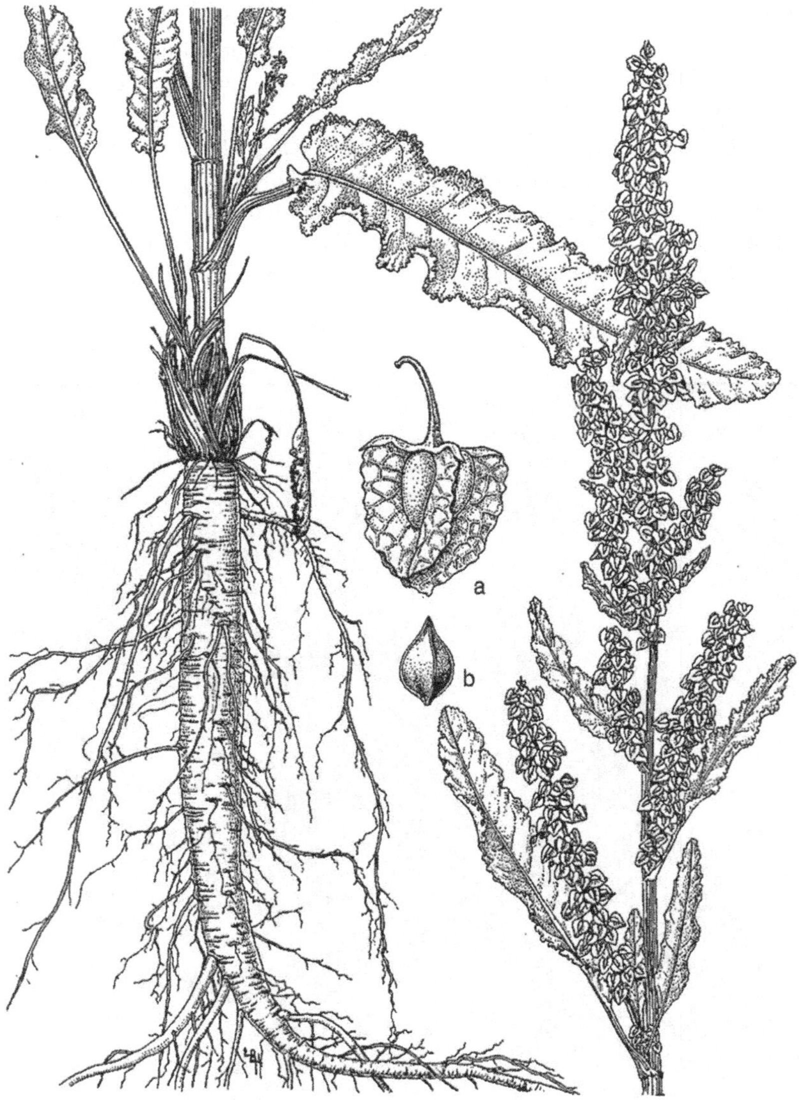

Fig. 9.264 Curly Dock (*Rumex crispus*)—POLYGONACEAE. Eighteen to fourty-eight inches tall, the fleshy stems die and regrow each spring. **Leaves:** Blue-green, alternate, 3–12 in. long with wavy edges. **Flowers**: Small green flowers produce red seeds in a large cluster 6–24 in. long that brown as they age. The flowers are not as dense as Canaigre's. **Seeds:** The flower's inner parts become large heartshaped wings 3–6 mm long and surrounding the tiny fruit. **a.** Fruiting calyx. **b.** Achene (Fig. 9.264 from *An Illustrated Guide to Desert Weeds*, by Kittie F. Parker. © 1972 The Arizona Board of Regents. Reprinted by permission of the University of Arizona Press)

Curly Dock (*Rumex crispus*) [P|M]

Except for its curly leaf margins, this Eurasian and North African perennial resembles its American relative (Canaigre). At CWF, it grows in the same moist habitats, depressions, pond and river edges, and lawn margins (Fig. 9.265).

Curly Dock has medicinal value. You can even buy it transformed into small round pills. Use Moore's 2003 book to learn more.

Fig. 9.265 Curly Dock seeds (photo © Stickpen)

9.30 Purslane Family: PORTULACACEAE (Fig. 9.266)

Fig. 9.266 Common Purslane (*Portulaca oleracea*)—PORTULACACEAE. Smooth fleshy leaves on reddish prostrate branches 6–24 in. long. Can form mats with tips turned up. Single plants can be upright to about 12 in. **Leaves**: *The leaves are alternate not opposite as in similar Horse Purslane (Trianthema portulacastrum, AIZOACEAE family) a bitter tasting plant, and Woolly Tidestromia (Tidestromia lanuginosa) a hairy plant which I have not tasted.* The leaves are spatulate with rounded tips, and grow singly or in clusters. Horse Purslane leaves are round. **Flowers**: Small stalkless flowers yellow, not purple like Horse Purslane, appear from April to June and are single or clustered in the leaf and branch axils and at stem tips. They open in the morning revealing 7–20 stamens. **Seeds**: The upper half of the globular seedpods falls when seeds are mature. The black flattened seeds are less than 1 mm long with a white spot at the scar. **a**. Flower. **b**. Seed (Fig. 9.266 from *An Illustrated Guide to Desert Weeds*, by Kittie F. Parker. © 1972 The Arizona Board of Regents. Reprinted by permission of the University of Arizona Press)

Common Purslane (*Portulaca oleracea*) [A|DEZ-1]

This European annual has been with us a long time. As with Common Lambsquarters, Purslane cultivation has gone on for thousands of years.

Purslane occurs throughout the Old World. Its means of reaching the New World is unknown, but evidence shows it was present during pre-Columbian times from 1350 to 1539.

Present in many prehistoric sites, the plant appears in historical records from the seventh century BC. Theophrastus advised sowing Purslane in April, and Pliny the Elder recommended wearing a Purslane amulet to block evil.

Purslane contains more omega-3 fatty acids (alpha-linolenic acid in particular) than any other leafy vegetable. Moreover, one cup of cooked leaves contains 561 mg of potassium, 90 mg of calcium, and 2000 units of vitamin A. Wikipedia describes harvesting and preparing plants for lunch (or dinner).

Despite its many benefits, weed scientists consider Purslane to be one of the ten most noxious weeds worldwide (Fig. 9.267.

Fig. 9.267 Common Purslane (photo © GR)

9.31 Pondweed Family: POTAMOGETONACEAE (Fig. 9.268)

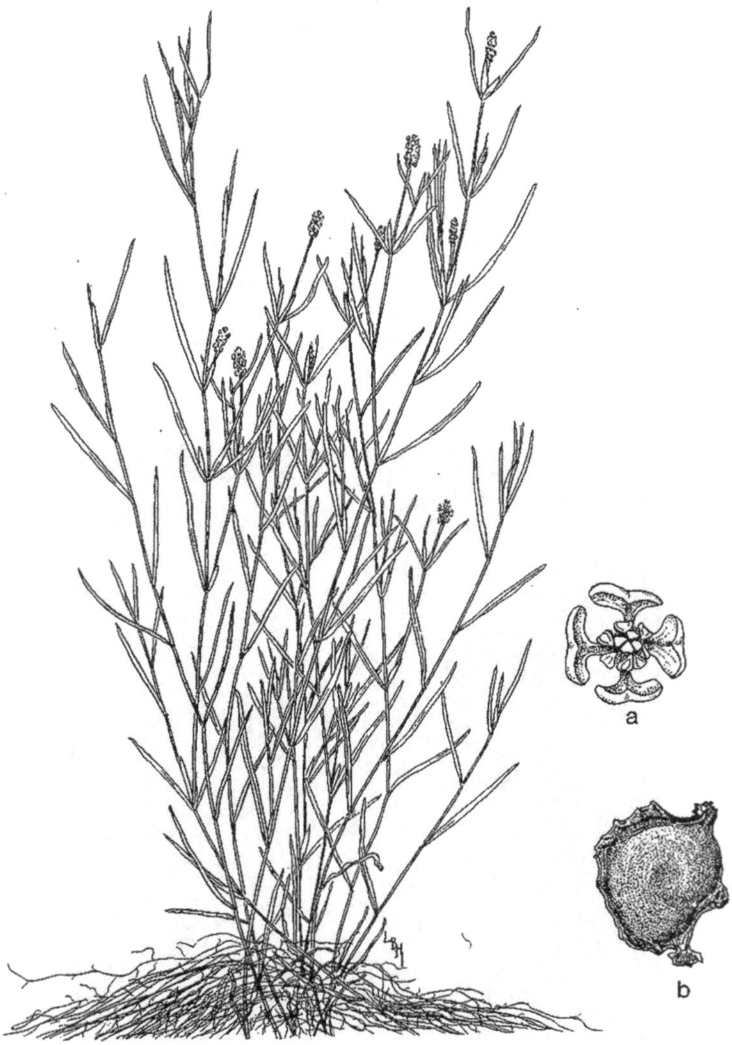

Fig. 9.268 Leafy Pondweed (*Potamogeton foliosus*)—POTAMOGETONACEAE. Aquatic perennial reproduces by seed and from rhizomes. Threadlike branching stems as long as 42 in. **Leaves**: Alternate, submerged, deep green to bronze, to 3 mm wide and 4 in.. long. **Flowers**: Tiny, stalkless with four brown petallike parts 1 mm long clusters of 2–12 flowers above water while in flower but submerged by the time fruit matures. **Seeds**: Nutlets 2–3 mm long, greenish brown. **a**. Flower with four petallike parts. **b**. Nutlet showing the typical beak and scalloped keel (Fig. 9.268 from *An Illustrated Guide to Desert Weeds*, by Kittie F. Parker. © 1972 The Arizona Board of Regents. Reprinted by permission of the University of Arizona Press)

Leafy Pondweed (*Potamogeton foliosus*) [P|N]

This weed is a valuable source of food and cover for aquatic wildlife. We can lose it when nitrates from farms and human wastes promote heavy growth by floating algae. The algae block light and deplete oxygen. A few years after I moved to the farm, the algae became so dense in one pond it killed off most of the plants and all of the fish.

Fifteen years ago, I added Grass Carp, Koi, and Goldfish fingerlings to the ponds at CWF to help control the algae. The fish did well. In 10 years, the Koi exceeded 20 in in length. They reproduced and together with the Carp and Goldfish, they were doing a fine job controlling the algae. Ten years ago, a Bald Eagle began fishing the ponds. Five years ago, a pair of Double-crested Cormorants joined the Eagle. In 2017, I saw two Koi and no Goldfish.

Leafy Pondweed can behave badly in warm climates. It can form mats so dense they block the flow of water in irrigation canals (Fig. 9.269).

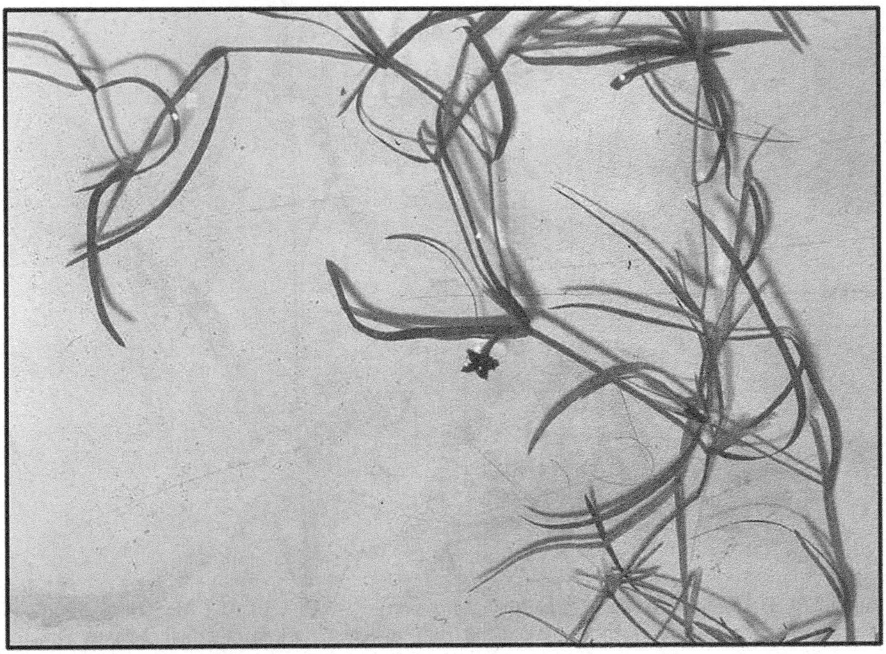

Fig. 9.269 Leafy Pondweed (photo by Mohlenbrock 1995; USDA)

9.32 Buttercup Family: RANUNCULACEAE (Fig. 9.270)

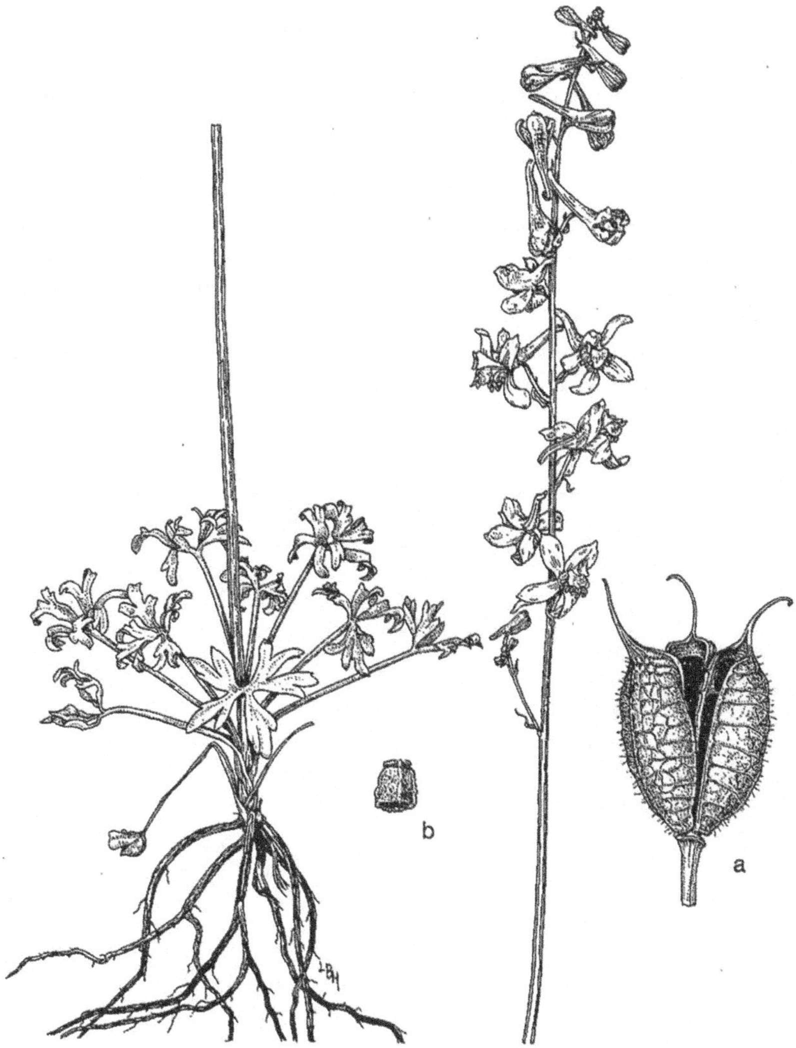

Fig. 9.270 Barestem Larkspur (*Delphinium scaposum*)–RANUNCULACEAE. This 6–30 in. tall gray-green perennial reproduces from seed and from clusters of dark woody roots. **Leaves:** Divided into narrow segments. **Flowers:** Vivid royal blue sepals with long basal spurs. **a.** Seedpod (1–2 cm long). **b.** Seed (3 mm) (Fig. 9.270 from *An Illustrated Guide to Desert Weeds*, by Kittie F. Parker. © 1972 The Arizona Board of Regents. Reprinted by permission of the University of Arizona Press)

Barestem Larkspur (*Delphinium scaposum*) [P|DMN]

This drought-tolerant native plant rises from seeds and perennial roots. New shoots form patches that flower spring through summer. My colonies have lasted 2–3 years (Fig. 9.271).

The plant's foliage poisons cattle, but the flowers are fabulous insect magnets. Butterflies, bees, flies, and moths of many species like the nectar. Around my house, Larkspur is the most popular flower for bumblebees including the big shiny black carpenters.

Moore (2003) describes preparation and use of Larkspur seeds to control lice. Good to know.

Fig. 9.271 Royal blue to purple Barestem Larkspur is a common desert weed (photo © GR)

9.33 Snapdragon Family: SCROPHULARIACEAE (Fig. 9.272)

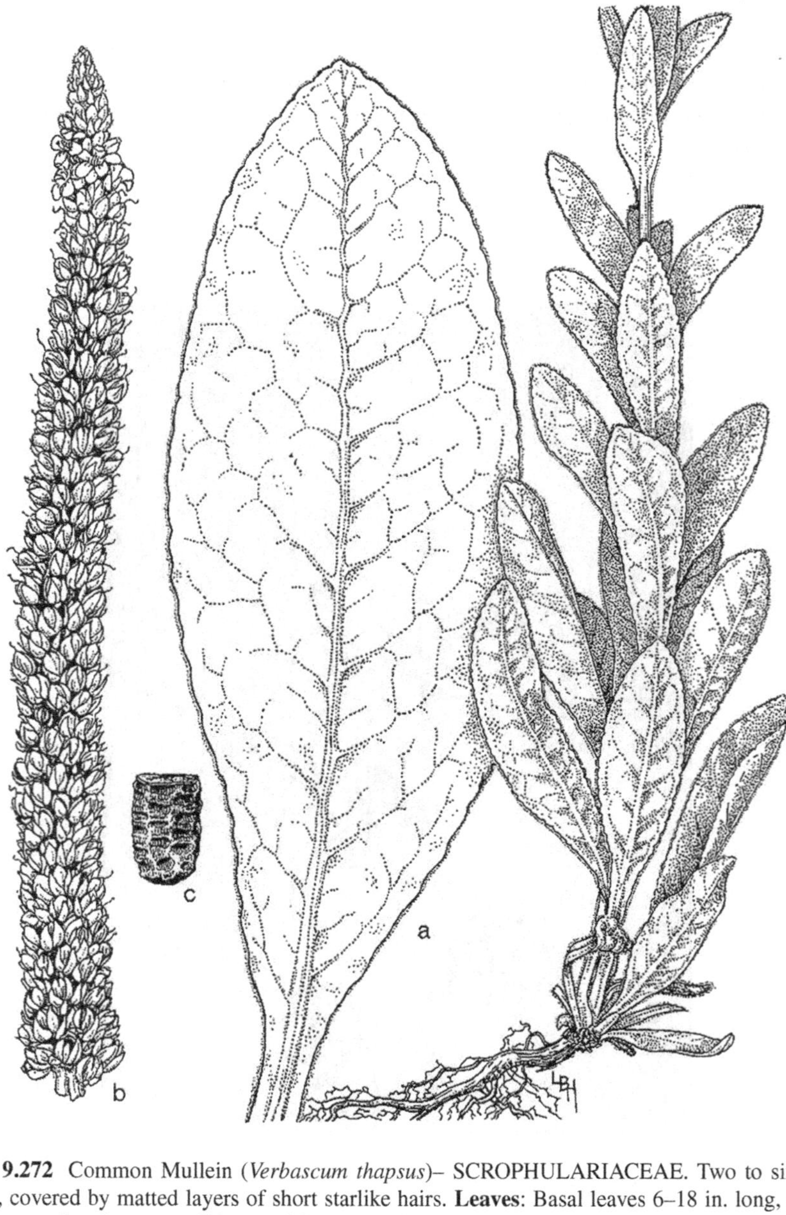

Fig. 9.272 Common Mullein (*Verbascum thapsus*)– SCROPHULARIACEAE. Two to six feet high, covered by matted layers of short starlike hairs. **Leaves**: Basal leaves 6–18 in. long, upper leaves smaller. Leaves stalkless and bases attach to the stem down to the next leaf. **Flowers:** Lemon yellow, stalkless, five-lobed, 20–25 mm across, crowded on a 1–3 ft. spike at the top of the stem. The spikes are 20–32 mm thick. **Seeds**: Tomentose eggshaped seedpods about 6 mm in diameter contain innumerable dark brown seeds less than 1 mm long. **a**. Woolly basal leaf. **b**. Upper portion of fruiting spike with flowers at tip and seedpods below. **c**. Seed, enlarged (Fig. 9.272 from *An Illustrated Guide to Desert Weeds*, by Kittie F. Parker. © 1972 The Arizona Board of Regents. Reprinted by permission of the University of Arizona Press)

Common Mullein (*Verbascum thapsus*) [B|M]

Big thick woolly leaves and stout flower spikes make Mullein unmistakable. This European biennial begins life with a basal rosette the first year and the tall stem the second year. Flowers appear from June to August and sometimes later. The plant colonizes disturbed soil along roads, railroads, transmission corridors, and overgrazed ranges (Fig. 9.273).

Mullein is a safe remedy for many medical problems, and it has other uses. "Although not comparable in luxury to two-ply, scented bathroom tissue, the leaves are sort of floral designed and may be used similarly—and there is no chance of confusing it with poison oak" (Moore 2003, p. 173).

Fig. 9.273 Mullein outstanding in its field (© GR)

9.34 Quassia Family: SIMAROUBACEAE (Fig. 9.274)

Fig. 9.274 Tree of Heaven (*Ailanthus altissima*)—SIMAROUBACEAE. Tree to 20 m often producing clonal thickets from rhizomes. **Leaves**: Pinnately compound 30–100 mm long with 9–31 lanceolate leaflets 5–15 cm long with a *single terminal leaflet*. (Arizona Walnut leaves are similar but smaller: 18–38 cm long with 9–15 leaflets. They have a small terminal leaflet or none.) **Flowers**: In large terminal panicles 10–40 cm long with small green to white flowers with woolly petals 2–3 mm long. **Fruit**: Samaras (winged achenes) 3–5 cm long, twisted one-half turn at the top (Fig. 9.274 photo © GR)

Tree of Heaven (*Ailanthus altissima*) [P|IMT]

This common tree forms dense patches on roadsides and in yards. It tolerates harsh conditions, which could be good, but it is malodorous and spreads as fast as zombies did in the movie *World War Z* (Carnahan et al. 2013). The tree does grow throughout New York City where I have seen it doing fine in dark slots between apartment buildings. Eradication of established clones requires initial sawing or chopping followed by vigilance to eliminate sprouts before they make leaves and contribute energy to the roots (Fig. 9.275).

Traditional Chinese medicine uses Tree of Heaven. Wikipedia gives a detailed example of a cure for mental illness that uses the tree's roots, urine from young boys, and a black-bean concoction. There are other, less exotic sounding medicines described in Wikipedia, but Moore (1989, 2003) does not mention Tree of Heaven.

Fig. 9.275 Tree of Heaven leaves and samaras (photo © GR)

9.35 Potato Family: SOLIANACEAE (Fig. 9.276)

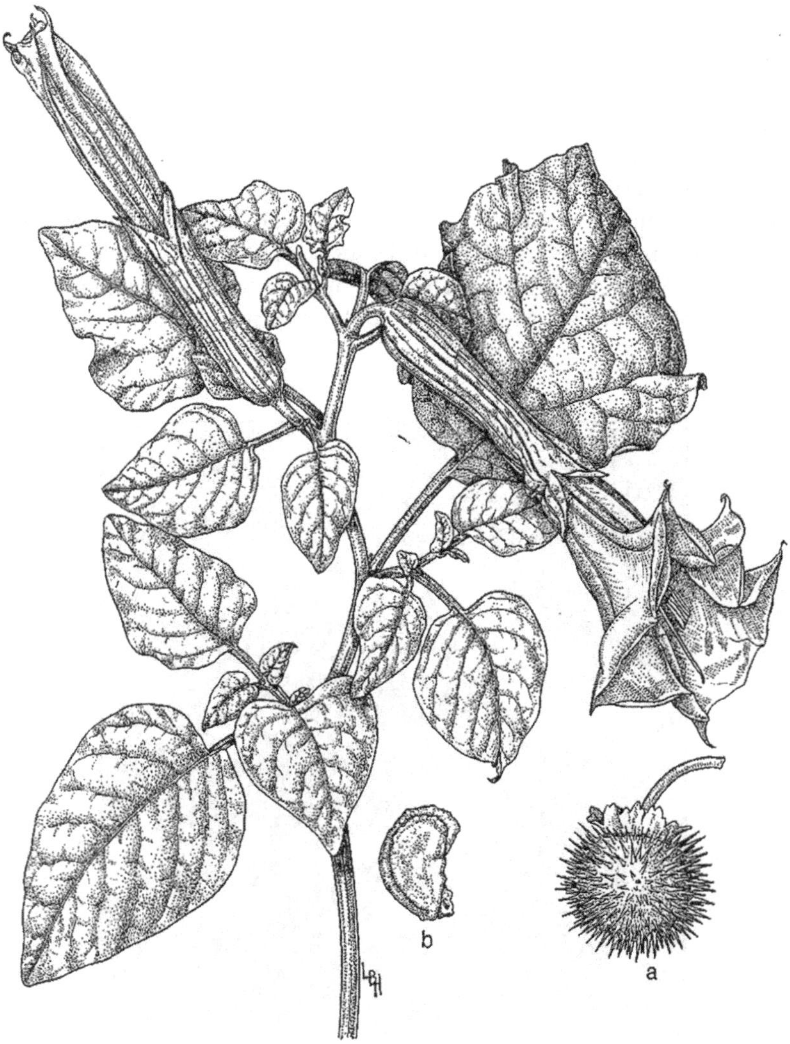

Fig. 9.276 Sacred Datura (*Datura innoxia*)—SOLIANACEAE. Large gray-green perennial 2–3 ft. tall forms spreading clumps. **Leaves**: Alternate, eggshaped, 3–10 in. long with wavy toothed edges and pointed tip. White veins obvious especially below. **Flowers**: Fragrant, large, white or pale lavender, funnelshaped, *6–10 in. wide* with slender teeth 1–2 cm long. Arise in forks of stems, open in the evening and close by noon. **Fruit**: Hard round seedpods with 1 cm long spines contain yellow-brown seeds. **a**. Spiny seedpod. **b**. Seed (Fig. 9.276from *An Illustrated Guide to Desert Weeds*, by Kittie F. Parker. © 1972 The Arizona Board of Regents. Reprinted by permission of the University of Arizona Press)

Sacred Datura (*Datura innoxia*) [P|DMN]

Carlos Castaneda popularized this native weed in a mystical memoir published in 1968. Adventurous youngsters are still getting their stomachs pumped. Datura has several medicinal uses, but "This plant should never be taken internally for any reason" (Moore 2003) (Fig. 9.277).

Fig. 9.277 Sacred Datura—common across southern North America and scattered in South America (photo © GR)

Fig. 9.278 Woodland Tobacco (*Nicotiana sylvestris*)—SOLIANACEA. Annual herb to 66 in. tall. **Leaves** lightly sticky, partially clasping. Largest leaves at base to 2 ft. long and 10 wide. **Flowers** in racemes, tubular, to 8 cm long with 32 mm wide bell. Powerful sweet scent strongest at night. Tiny **seeds** about 1 mm, brown (Fig. 9.278 photo © GR)

Another species, Desert Tobacco (*Nicotiana obtusifolia*) grows less than 1 m tall with leaves to 15 cm long and with small flowers to 1 cm long. The plant's leaves have glandular sticky hairs that give it a bitter scent.

Woodland Tobacco (*Nicotiana sylvestris*) [A|DEM]

This South American plant has interesting flowers with sweet fragrance that intensifies after dark to attract nocturnal pollinators. The nursery industry disperses Tree Tobacco as an ornamental. Online sources say the plant is perennial, but I see no evidence of that at CWF. Perhaps in a warmer climate or in a very sheltered location it could overwinter.

Desert Tobacco is more widespread than Woodland Tobacco in North American deserts, but it is not present at CWF. I know of no instances of establishment of this or several other New World tobaccos on Old World continents. With its smaller flowers, nurseries are less likely to offer desert tobacco as an ornamental (Fig. 9.279).

Fig. 9.279 Woodland Tobacco (photo © GR)

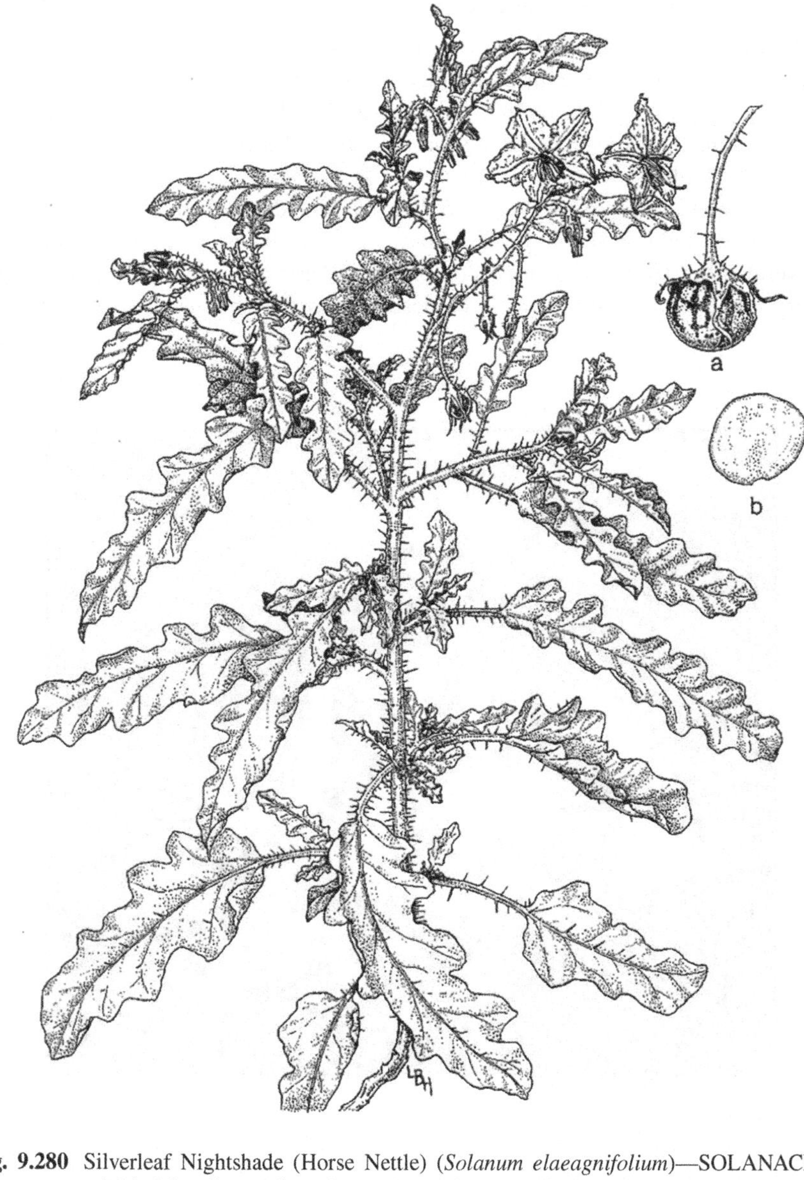

Fig. 9.280 Silverleaf Nightshade (Horse Nettle) (*Solanum elaeagnifolium*)—SOLANACEAE. Silvery perennial 1–3 ft. tall reproduces by seeds and rhizomes. Entire plant covered by matted tiny starlike hairs and slender yellow spines. The spines are sometimes absent. **Leaves**: Alternate, lanceshaped, 1–4 in. long, and 6–25 mm wide. Margins wavy. **Flowers**: Yellow stamens surrounded by deep violet or blue 19–25 mm wide petals. **Seeds**: **a.** berrylike hairless pods 8–12 mm in diameter mottled green, becoming yellow or orange contain many seeds. **b.** Seeds disk shaped, 3 mm long, yellow-brown, shiny, and finely granular (Fig. 9.280 from *An Illustrated Guide to Desert Weeds*, by Kittie F. Parker. © 1972 The Arizona Board of Regents. Reprinted by permission of the University of Arizona Press)

Silverleaf Nightshade (*Solanum elaeagnifolium*) [P|INM]

Like Crested Pricklepoppy, Silverleaf Nightshade's thorns and toxic sap make it cow proof. It flourishes in cow and horse pastures as the grazers remove the competition. Though this is a native weed, its name (*elaeagnifolium*) refers to its leaves' resemblance to introduced Russian Olive. The plant has spread around the world. Able to tolerate both arid and mesic conditions, Silverleaf Nightshade is a weed of national significance in Australia, and an invasive problem in many regions.

High alkaloid concentration makes this weed one of our most poisonous. Other members of the SOLANACEAE are also poisonous, but some are edible. Tomato, potato, and eggplant are familiar edibles. The family connection is clear if you compare potato and Silver Nightshade flowers (Fig. 9.281).

Silverleaf Nightshade's potent chemistry is a source of medicinal compounds. Moore (1989, 2003) does not mention the plant, but alkaloid use appears in literature cited on the website of the Southwest School of Botanical Medicine, Bisbee, Arizona (2019).

Fig. 9.281 Silverleaf Nightshade (photo © GR)

Fig. 9.282 Buffalobur (*Solanum rostratum*)—SOLANACEAE. Stems branching in upper part of plant are 6–24 in. long. Straight spines cover stems, leaf veins, and seedpods. Yellow flowers are five-lobed, 25–38 mm wide. The spiny calyx enlarges and encloses the seedpod in a spiny bur. **a**. Seeds are about 12 mm in diameter (Fig. 9.282 from *An Illustrated Guide to Desert Weeds*, by Kittie F. Parker. © 1972 The Arizona Board of Regents. Reprinted by permission of the University of Arizona Press)

Buffalobur (*Solanum rostratum*) [A|NP]

People probably transported this native annual of the Great Plains to Arizona where it has spread along roads into yards, irrigated farms, and across overgrazed land. It now covers thousands of acres.

Parker (1972) called Buffalobur vicious, and when you find one, you will agree. But more than appearing fiendish, the plant backs up its scary appearance with solanine, a glycoalkaloid poison in its leaves and stems.

Buffalobur became invasive when transported from its native home east of the Rocky Mountains. The plant requires livestock grazing or other disturbance to provide places for its seeds to germinate. Once established, it can hold its ground and prevent normal succession by other plants (Fig. 9.283).

Fig. 9.283 Buffalobur (photo © Jerry Friedman)

9.36 Tamarisk Family: TAMARICACEAE (Fig. 9.284)

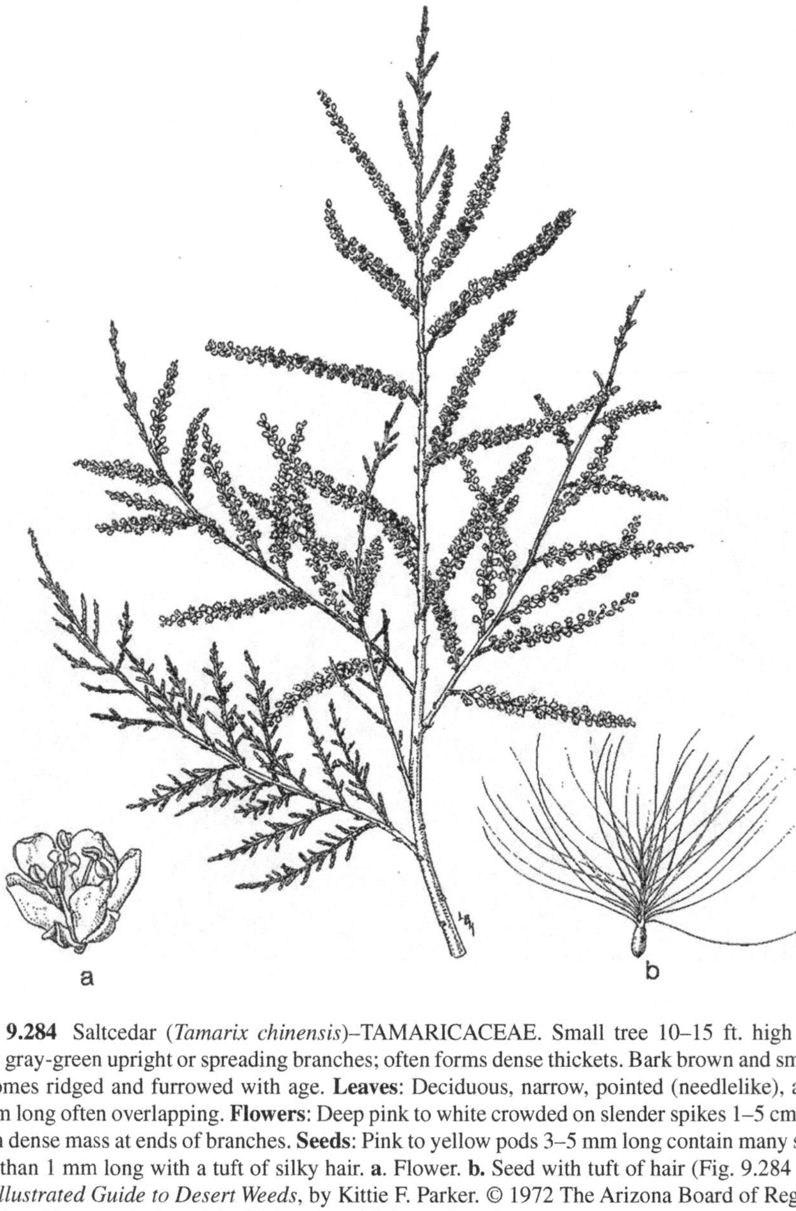

Fig. 9.284 Saltcedar (*Tamarix chinensis*)–TAMARICACEAE. Small tree 10–15 ft. high with long gray-green upright or spreading branches; often forms dense thickets. Bark brown and smooth becomes ridged and furrowed with age. **Leaves**: Deciduous, narrow, pointed (needlelike), about 2 mm long often overlapping. **Flowers**: Deep pink to white crowded on slender spikes 1–5 cm long form dense mass at ends of branches. **Seeds**: Pink to yellow pods 3–5 mm long contain many seeds less than 1 mm long with a tuft of silky hair. **a**. Flower. **b.** Seed with tuft of hair (Fig. 9.284 from *An Illustrated Guide to Desert Weeds*, by Kittie F. Parker. © 1972 The Arizona Board of Regents. Reprinted by permission of the University of Arizona Press)

Saltcedar (*Tamarix chinensis*) [P|IM]

(Five-stamen Tamarisk, Tamarack) For 2 years, my family lived on the Little Colorado River floodplain. Exploring the maze-like alleys and tunnels floored by inches of accumulated needles and roofed by arching branches gave me many hours of entertainment in the dense Tamarix woodland beside the river. This Eurasian native smothers most other plants that once populated the floodplain. It is present beside the Agua Fria River in CWF and in other moist locations around the farm. I cut all the clumps along the river when I moved in. They have recovered, and it is time to cut them again. Left alone, they would spread to defeat the willows, walnuts, and cottonwood sprouts. According to contributors to the IUCN Global Invasive Species Database, this is one of the 100 world's worst weeds.

Gardeners use Saltcedar as a landscape ornamental, and it provides pollen for bees and cover for birds. The endangered Southwest Willow Flycatcher nests in the branches. According to SEINet, the plant is also useful for curing camel mange. Of course it is. Nevertheless, Saltcedar forests are impoverished places where few species live. Sad a day only they remain along our desert streams (Fig. 9.285).

Fig. 9.285 Saltcedar flowers (photo © Arashiyama)

9.37 Cattail Family: TYPHACEAE (Fig. 9.286)

Fig. 9.286 Common Cattail (*Typha latifolia*)—TYPHACEAE. Four to seven feet tall. **Leaves**: Twelve to sixteen, flat 16–23 mm wide. **Flowers**: Usually no space between the lower (female or pistillate) and upper (male or staminate) flowers. Flowers packed into a cigar shape 4–7 in. long and 2–3 cm thick. Male flowers with 2–5 stamens fall as soon as the pollen sheds leaving the upper end of the spike naked. Female flowers, each with 40–60 fine hairs, persist for months. **Seeds**: Minute brown achenes 1–2 mm long with fine hairs still attached when shed. **a**. Plant with rhizomes. **Aa**. Flowering spike; staminate flowers above, pistillate flowers below. **Ab**. Achene of **b** (Southern Cattail), not a current CWF resident (Fig. 9.286 from *An Illustrated Guide to Desert Weeds*, by Kittie F. Parker. © 1972 The Arizona Board of Regents. Reprinted by permission of the University of Arizona Press)

Common Cattail (Bulrush) (*Typha latifolia*) [P|DEN]

Bulrushes are perennials that grow in freshwater marshes, along rivers, around ponds, and in irrigation canals. They were present in North and South America, Africa, and Eurasia when recorded history began. Since then, they have spread to Hawaii, Australia, and New Zealand. The plants reproduce from seed and rhizomes. They bloom from early to late summer.

Shoots from the rhizomes form dense stands difficult to eradicate. Seasonal fluctuation of water level in ponds and desert streams, especially those that dry out, can create enough stress to prevent the plants from spreading. Elsewhere, people have eradicated Bulrush by repeated cutting below the water surface to deny oxygen to the roots and rhizomes.

Bulrushes are important members of the cool deserts' moist ecosystems. They stabilize disturbed areas and provide cover for ducks and other wildlife. They provide food for muskrats, they host moths, beetles and other insects, and their fruits and stems provide material for bird nests (Fig. 9.287).

Fig. 9.287 Cattail in a CWF pond near the Agua Fria River (photo © GR)

9.38 Elm Family: ULMACEAE (Fig. 9.288)

Fig. 9.288 Siberian Elm (*Ulmus pumila*)—ULMACEAE. Tree 15–30 m tall, deciduous with gray to brown furrowed bark. **Leaves**: Blade lanceolate, 2–7 cm long, and 2–4 cm wide, pointed, and single-toothed (most other elms are doubly toothed—i.e. their teeth have teeth). **Flowers**: In clusters of 5–15. **Seeds**: Tan, round, winged samaras, 10–14 mm in diameter, notched tip (Fig. 9.288 photo © Tiarescott—TerraNova, le désert de Gobi)

Siberian Elm (*Ulmus pumila*) [P|DEI] (Fig. 9.289)

This is the principal shade tree for many people living in arid regions similar to CWF. It casts dense shade, it is drought tolerant, and with regular water, it can grow 100 ft. tall.

This desert tree is an introduced invasive species illustrating the Janus nature of many weeds that can be both good and bad depending on time and place.

As many children will attest, the buds, leaves, flowers, and seeds are edible (Fig. 9.289).

Fig. 9.289 Shiny toothed leaves (photo © Melburnian)

9.39 Verbena Family: VERBENACEAE (Fig. 9.290)

Fig. 9.290 Prostrate Vervain (*Verbena bracteata*) Verbenaceae—VERBENACEAE. Low spreading annual or perennial 10–50 cm tall usually less than 25 cm. **Leaves** opposite, pinnate, 1–6.5 cm long. **Flowers** tiny white, blue, or purple about 4 mm wide. **Fruit**: 4 nutlets 2–2.5 mm long (Fig. 9.290 photo © GR)

Prostrate Vervain (*Verbena bracteata*) [A-B-P|MN]

Prostrate Vervain appeared at CWF in 2014 and began spreading on the driveway gravel near the bird garden. It is doing its job covering exposed ground and it is doing it without producing irritating seeds or odor. It seems a low mat-forming plant such as this is as effective as gravel in making the path navigable so it gets to stay. Native throughout North America, Prostrate Vervain has dispersed to Europe and Japan. I do not know if it is invasive in those places, but of course, local weed managers should keep an eye on it (Fig. 9.291).

Fig. 9.291 Prostrate Vervain flowers (photos © GR)

9.40 Violet Family: VIOLACEAE (Fig. 9.292)

Fig. 9.292 Sweet Violet (*Viola odorata*)—VIOLACEAE. Perennial with long green aboveground *cord-like runners* (stolens) that have thin roots or rhizomes at nodes. **Leaves**: On stalks arising from stolens, 2–6 cm long, dense short hairs, rounded indented base, blunt tip. **Flowers**: White to deep blue, fragrant, 1–2 cm long; five separate petals, all differently shaped. The two lateral petals have a beard of hairs near the base, and the base of the lowest petal forms a short, rounded spur or sac. **Fruit**: Purple eggshaped capsule opens from top releasing ivory seeds 3–4 mm long. **Insets**: Lower left: Intact capsule. Lower right: Dissected flower (Fig. 9.292 painting: Masclef 1891-1893)

Sweet Violet (*Viola odorata*) [P|D]

My mother's favorite flower, this weed grew wild on the banks of 14-Mile Creek where we lived. I did not think of it as a weed, and when it appeared at the edge of my back lawn at CWF, I was delighted. Then, after several quiescent years, the stolens shot outward in a dramatic and successful attack on the neighboring Bermudagrass. However, this may not be a problem; Javelina found the plant during the winter and began eating the new roots and shoots. Deer like the leaves (Fig. 9.293).

Fig. 9.293 Sweet Violet invasion into the lawn from the original patch at left. The new plants are more yellow than the older plants (photo © GR)

9.41 Aloe Family: XANTHORRHOEACEAE (Fig. 9.294)

Fig. 9.294 Onionweed (*Asphodelus fistulosus*)—XANTHORRHOEACEAE. Annual or short-lived perennial 20–70 cm tall forms tubers at the stem base. **Leaves**: Hollow tubes with hairy edges. **Flowers**: Close at night. Racemes to open panicles on 15–70 cm long branches. White to pale pink with dark pink or brown veins. Tepals (petals as in tulips and daffodils) 5–12 mm long and 2–3 mm wide. **Capsules**: Round 5–7 mm long with six seeds (Fig. 9.294photo © 2008 Hans Hillewaert)

Onionweed (*Asphodelus fistulosus*) [A|FI]

Livestock do not eat this Mediterranean weed. It is present in the Southwestern United States, Mexico, and the Middle East. It invades and spreads across rangelands and other disturbed areas with winter moisture and summer warmth.

I have not seen Onionweed in CWF. It grows in similar habitats and will probably reach us soon (Fig. 9.295).

Fig. 9.295 Onionweed (photo © Shachar haCaspi)

9.42 Caltrop Family: ZYGOPHYLLACEAE (Fig. 9.296)

Fig. 9.296 Bullhead (Puncturevine) (*Tribulus terrestris*)—ZYGOPHYLLACEAE. Spreading stems can form dense mats up to 10 feet in diameter. **Leaves**: Opposite, *pinnately compound* with 4–7 pairs of oblong leaflets 3–13 mm long. **Flowers**: Bright yellow on short stalks have five petals open in morning. Similar to Creeping Woodsorrel (above). **Seedpods**: Cluster of five nutlets each with two vicious spines. Each nutlet contains 2–5 seeds capable of remaining viable for many years awaiting sufficient moisture for germination. **a**. Flower. **b**. Seedpod with five bony nutlets. **c**. Nutlet (Fig. 9.296 from *An Illustrated Guide to Desert Weeds*, by Kittie F. Parker. © 1972 The Arizona Board of Regents. Reprinted by permission of the University of Arizona Press)

Bullhead (Puncturevine) (*Tribulus terrestris*) [A|MPZ-1]

Bullhead's origin is uncertain, but Eurasia is a safe bet. It germinates wherever seeds land, but seems to be most common in vacant lots, dry yards, street edges, and sidewalk cracks—all those places barefoot children walk or ride bikes with air-filled tires. On sunny mornings in mid to late summer, tiny yellow flowers expose Bullhead presence. The flowers resemble those of Creeping Woodsorrel (above). Check the leaves. Bullheads are pinnately compound and Woodsorrels are more like clover. Stop and pull the plants. Left to mature, they produce seeds with spines so long, hard, and sharp they puncture bicycle tires and leave dogs and barefoot children yelping and hobbling in pain. Most vexing, seeds stuck to shoe soles detach in carpet where they lie waiting for opportunities to animate the shoeless.

Two weevil species (*Microlarinus* spp.) native to France, Italy, and India can reduce Bullhead numbers. The weevils are available from biological supply companies, but they often do not survive long enough to eliminate all the plants. The best solution is to pull the plants before they drop any seed. Keep this up for a few years and the problem fades away. The plant is a regulated pest in Arizona. This means viable plant parts (stolons, rhizomes, cuttings, and seed, except agricultural, vegetable, and ornamental seed for planting purposes) found within the state may be controlled to prevent further infestation or contamination (Fig. 9.297).

Fig. 9.297 Bullhead patch of pain (photo © Forest and Kim Starr)

Chapter 10
Conclusion

I wrote this book with four questions in mind:

1. Do desert weeds differ from weeds of moist climates?
2. Which weeds are present on CWF, the 20-acre farm where I live? Answering this question requires identifying the weeds. Illustrations and descriptions of weeds fill most of the pages. These weeds provide examples of desert and dryland weeds worldwide. Their descriptions contain answers to the third question:
3. Are any desert weeds a valuable resource for wildlife and people?
4. How will climate change and the impacts of the growing human population affect weeds? Will weeds become more abundant?

Answers:

1. Weeds everywhere share the same basic characteristics. They have short lives, reproduce copiously, and establish in disturb sites. Fuelweeds, the species effective in carrying and surviving fire are common in deserts.
2. All of the 149 weed species in this book range far beyond CWF. In fact, people have introduced most of the 51 native weed species in the group to other continents where a few have become invasive.
3. All weeds are valuable in their natural role as protectors of the soil of disturbed areas. However, as people have spread them between continents, some become invasive and cause irreparable harm. In deserts and surrounding dry lands, competitive mechanisms and their flammability have allowed invasive weeds to spread and replace native plants and non-invasive weeds.

   The destruction of Earth's native ecosystems by invasive species is not near its end. And should people ever stop disturbing vegetation and dispersing weeds, it would be a long time before evolution could achieve a new equilibrium of interacting species. Considering the changes after the mass extinctions of the past, it is reasonable to expect the return of prehuman biological diversity will take thousands, perhaps even millions of years if it occurs at all.

© Springer Nature Switzerland AG 2020

G. Rogers, *Desert Weeds*, https://doi.org/10.1007/978-3-030-45854-6_10

4. And finally, the disruptive impacts of climate change, rising temperature, human overpopulation, and human overuse of natural resources will continue to disturb natural ecosystems and let weeds spread.

As human impacts and threats to Earth ecosystems have increased, the need to preserve species has grown. Scientists, concerned citizens, and concerned governments are racing to place seeds of many plants including many weeds in secure storage facilities around the world. For example, the Svalbard Global Seed Vault contains over one million sets of seeds (CropTrust 2019). You can search for a plant across all the global storage facilities online at the Genesys website. Not everyone will understand it is necessary to store plants in underground bunkers, but everyone knows we must plan for the worst while hoping for the best.

Afterword

The strip of urban wilderness where my boys and I walked is long gone. That reach of the Salt River became Tempe Town Lake bordered by luxury homes and businesses. Perhaps in some distant time, the dams will fail, the lakes will drain, the river will return, and its flood plain will host weeds and passing families again.

© Springer Nature Switzerland AG 2020

G. Rogers, *Desert Weeds*, https://doi.org/10.1007/978-3-030-45854-6

Internet Resources

A Note about Internet Sources

Internet information is valuable for its accessibility, but it is not stable. Internet publishers often delete or change the addresses of reports, comments, maps, and data. If a link fails to connect, it might be possible to relocate an item by accessing the root directory of the site. For example, if the address for lists of weeds defined according to the Federal Noxious Weed Act of 1974, http://plants.usda.gov/java/noxiousDriver, changes it might be helpful to access http://plants.usda.gov or even http://usda.gov and search for the new location.

I cited other sources of information in the text and included them in the Sources list above.

Global Register of Introduced and Invasive Species (2019) http://www.griis.org. Accessed 6 Aug 2019.

Heap I (2018) The International Survey of Herbicide Resistant Weeds www.weedscience.com. Accessed 6 Aug 2019.

International Union for the Conservation of Nature (IUCN), Invasive Species Specialist Group (2019) http://www.issg.org/worst100_species.html. Accessed 6 Aug 2019.

North American Invasive Species Management Association (2019) http://www.naisma.org. Accessed 6 Aug 2019.

Play, Clean, Go—Stop invasive species in your tracks (2019) http://www.playcleango.org. Accessed 6 Aug 2019.

Society for Ecological Restoration (2019) https://www.ser.org. Accessed 6 Aug 2019.

U. S. National Invasive Species Information Center https://www.invasivespeciesinfo.gov. Accessed 28 June 2019.

Weed Science Society of America (2019) Publishes: *Invasive Plant Science & Management, Weed Science, Weed Technology*. http://wssa.net. Accessed 6 Aug 2019.

© Springer Nature Switzerland AG 2020

G. Rogers, *Desert Weeds*, https://doi.org/10.1007/978-3-030-45854-6

Glossary

Botanical terminology has a history stretching back thousands of years. Some of the terms we use to label plant parts are unique to botany, but most of them are forms of common adjectives.

This glossary is limited to terms used in this book. For a more complete list, I recommend the Glossary of Botanical Terms on Wikipedia and the fine book by David Gledhill (Gledhill 2008). The definitions below are from several sources, but I tended to follow Parker's plain style (Parker 1972). Wikipedia has illustrations for most of them.

Achene Small dry one-seeded fruit in which the ovary wall is free from the seed
Acuminate Gradually tapering to a sharp point
Acute Sharp pointed but less tapering than acuminate; angle less than 90 degrees
Allelopathic Produces biochemicals that prevent the germination, growth, survival, or reproduction of neighboring plants
Alternate Single leaves attached at a node
Annual Plant that usually germinates, flowers, and dies in one growing season
Anther Enlarged part at the top of the stamen; bears the pollen
Attenuate Gradually narrowing to a slender apex or base
Auricle Some grasses have small claw-like extensions of the collar that separates the sheath from the blade
Awn More or less stiff bristle on the end of lemmas (lower bracts) or glumes (paired bracts)
Axil Upper surface of angle between the stem and the leaf
Blade Grass leaf above the sheath
Bract Modified leaf underneath a flower or inflorescence
Bristle Stiff, slender appendage
Bunchgrass Grass without stolons or rhizomes; growth forms a bunch
Bur Rough or prickly envelope around a seed or fruit
Calyx Outer set of floral leaves, usually green, composed of all the sepals
Campanulate Bell-shaped.

© Springer Nature Switzerland AG 2020
G. Rogers, *Desert Weeds*, https://doi.org/10.1007/978-3-030-45854-6

Caryopsis Dry one-seeded fruit in which the ovary wall unites with the seed coat, typical of grasses
Ciliate Fringed with hairs on edge
Collar Region of the outer side of the grass leaf at the junction of the blade and sheath
Compound Leaves divided into leaflets
Convolute Rolled longitudinally
Corolla All the flower petals
Corymb (adj. corymbiform) is a flower array or cluster with outermost flowers on longer pedicels than the inner
Creeping Spreading just under the surface of the soil
Culm Stem of grass plants
Cyathia Euphorbia flowers appear to be individual flowers, but they are composed of bracts fused into a cup (involucre), with peripheral nectary glands supported by petal-like (petaloid) bracts. Within the cup, there is a ring of male flowers, each with a single stamen. Out of the middle protrudes a single, stalked female flower with no petals
Deciduous Sheds leaves at the end of the growing season. Induced by cold temperature or by drought
Decumbent Stems that lie flat and turn up at their ends
Decurrent Extending downward from the point of attachment
Dehiscent Splitting open at maturity
Dioecious Plants with only male (staminate) flowers
Disk flowers Tubular flowers in the center of the flower heads in the aster/sunflower family
Distally Opposite point of attachment
Embryo Young plant enclosed in a seed
Entire Continuous smooth leaf margin
Entity Natural vegetation unit defined by particular plant species
Erose Irregularly notched at the apex; appearing gnawed or eroded
Flexuous Bent alternately in opposite directions; a wavy form
Floret Individual grass flower including its two bracts, lemma and palea, or flower heads in the sunflower family
Floret Small flower in the spikelet of the grasses, or the flower head of the sunflower family
Flower head Dense inflorescence of sessile (stalkless) flowers
Fruit Seed case or seed pod (the ripened ovary); containing one to many seeds
Geniculate Bent sharply, like a knee
Glabrous Smooth without hairs
Glands Secreting tissue
Glaucous Powdery or waxy surface cover
Glumes Two thin bracts surrounding the spikelet of a grass (forming the husk of a cereal grain); or one surrounding the florets of a sedge
Grain The seed-like dispersal unit of the grass family
Hirsute With straight rather stiff hairs

Hispid With long stiff hairs
Hyaline Colorless and translucent
Imperfect Flowers with either pistils or stamens; unisexual
Indehiscent Not opening at maturity
Inferior ovary Sepals, petals, and stamens appear to rise from the very top of the ovary
Inflorescence Flowering part of a plant
Internode Part of a stem between two successive nodes
Introduced Species people brought to North America
Invader Undesirable plants that invade and replace native vegetation
Involucre Cluster of bracts below the inflorescence or cone (involucrate = having an involucre)
Involute Rolled inward on the upper (dorsal) surface
Keel Central dorsal ridge; the two united petals in some legume flowers
Knotty Hardened mass at the base or nodes
Lacerate Appearing torn at the edge or irregularly cleft
Leaflet Division of a compound leaf
Lemma Lower bract of the two bracts of a grass floret
Ligule A thin lining on the inner side of the leaf sheath at its junction with grass and sedge blades
Ligule the membranous, toothed, or hairy appendage on the inside of a leaf exposed at the junction of the sheath and blade
Membranous Thin, like a papery membrane
Monocarpic Flowering and bearing fruit once and then dying; can apply to annuals, biennials, or perennials
Monoecious Plants with only female (carpellate) flowers
Mycorrhiza Formed by a combination of underground threads usually associated with plant roots
Native of Species that originated in North America
Nerve Vein on glume, lemma, or palea
Node Place on a stem where leaves and branches originate
Nutlet Small nut; the one-seeded portion of a larger fruit that separates at maturity
Obtuse blunt or rounded (apex or base), usually making an angle of more than 90 degrees
Opposite Leaf pairs together one on each side of the stem
Ovary Base of the female part of the flower (pistil); ripens into the fruit
Palea Innermost of the two bracts of a grass floret
Panicle Loose, irregularly branched inflorescence with stalked individual flowers
Pappus (plural pappi) Modified calyx of a disk or rayflower in the sunflower family; consists of awns, scales, or bristles at the apex of the achene. Facilitates wind dispersal
Pedicel Stalk of a spikelet
Perennial Plant living more than 1 year
Perfect flowers Both male and female reproductive parts in the same structure
Perianth Outer part of flowers, consisting of the calyx (sepals) and corolla (petals)

Petaloid Petallike
Phyllary In asters, a bract that with others forms the involucre or cup that holds the flowers
Pilose Covered with soft hair
Pinnate Leaf divided into leaflets along a central stem
Pistil Female reproductive structures of a flower
Polycarpic Produces flowers or spores more than once during lifetime
Pubescent Covered with soft hairs
Raceme A flower cluster with the separate flowers on short equal stalks at equal distances along a central stem (rachis). The flowers at the base of the central stem develop first
Ray flower Marginal petal flowers in the inflorescence of the aster/sunflower family
Retrorse Directed back or downward
Rhizomatous Having rhizomes
Rhizome Horizontal underground stem
Rosette Cluster of spreading or radiating basal leaves
Rotate Wheel-shaped usually applied to a circular, nearly flat, short-tubed, sympetalous corolla.
Runner Horizontal aboveground stem; a stolon
Samara Winged achene
Scabrous Roughened with stout projections
Scurf (adj. scurfy) Small branlike scales on the surface of a leaf or other part.
Sepal One of the outer set of floral leaves forming the calyx, usually green
Sessile Stalkless leaves or flowers
Sheath In grasses, the basal portion of the leaf forms a tube (sheath) wrapping around the stem
Silicles Dry, dehiscent fruit of the Brassicaceae with two parts separating at maturity; fruit is typically *less than three times as long as wide*
Silique Dry, dehiscent fruit of the Brassicaceae with two parts that separate at maturity; fruit is typically *more than three times as long as wide*
Sinuate, sinuous (adv. Sinuately) Having a deep wavy margin.
Sod-grass Grass with stolons or rhizomes; may form continuous thatched surface
Spatulate Narrow at the stem widening to the tip
Spike An unbranched inflorescence with stalkless flowers
Spikelet the part of a grass inflorescence consisting of two glumes and one or more florets
Stamen The male portion of the flower
Stipules Pair of tongue-like appendages at the base of the leaf
Stolon Horizontal stem that roots along its length to form new plants; aboveground or at very shallow depth
Stoloniferous Bearing stolons
Striate Marked with slender, longitudinal grooves or lines, appearing striped
Strigose Stiff, straight hairs laying on a surface
Succulent Soft and fleshy in texture

Tomentose Covered by dense matted and tangled hairs, velvety
Truncate "Cut off" at the base or apex
Truncate-ciliate Ligule with short fringe of hair tips
Tuber A swollen underground stem
Umbel Flower stalks arising from a common point a bit like an umbrella blown inside out.
Utricle Small thin-walled fruit with a covering looser and more fragile than that of an achene. Common among Amaranths. The Tumble Pigweed drawing and one of the photos illustrate the utricle.
Villous Densely hairy with long, soft hairs
Whorl Three or more leaves or flowers arising from the same node

References

Abler R, Adams JS, Gould P (1971) Spatial organization: the geographer's view of the world. Englewood Cliffs, Prentice Hall

Arizona Regulated and Restricted Noxious Weeds 2019 State of Arizona. https://agriculture.az.gov/pests-pest-control/agriculture-pests/noxious-weeds. Accessed 26 June 2019

Babbitt B (1998) Statement by secretary of the interior on invasive alien species. In: Science in wildland weed management. U. S. Department of the Interior, Denver

Beckman NG, Bullock JM, Salguero-Gomez R (2018) High dispersal ability is related to fast life-history strategies. J Ecol 106:1349–1362

Benbrook CM (2016) Trends in glyphosate herbicide use in the United States and globally. Environ Sci Eur. https://doi.org/10.1186/s12302-016-0070-0

Bennett PS, Kunzmann MR, Graham LA (2004) Descriptions of Arizona vegetation represented on the GAP vegetation map. Biological Resources Division, U. S. Geological Survey, Denver

Billings WD (1990) Bromus tectorum, a biotic cause of ecosystem impoverishment in the Great Basin. In: Woodell GM (ed) The earth in transition: patterns and processes of biotic impoverishment. Cambridge University Press, Cambridge, pp 301–322

BLM (2007) The Bureau of Land Management's performance and accountability report for fiscal year. U. S. Department of the Interior, Bureau of Land Management, Washington: Government Printing Office, Washington

Brodie G (2018) The use of physics in weed control. In: Jabran K, Chauhan BS (eds) Non-chemical, weed control. Academic Press, London. https://doi.org/10.1016/B978-0-12-809881-3.00003-6

Brown DE (1982) Biotic communities of the American Southwest—United States and Mexico. Desert Plants 4:1–342

Burcham LT (1970) Ecological significance of alien plants in California grassland. Proc Assoc Am Geogr 2:36–39

Burkhardt JW (1996) Herbivory in the intermountain west: an overview of evolutionary history, historic cultural impacts, and lessons from the past. Station Bulletin 58. Idaho Forest, Wildlife and Range Experiment Station, Univ. of Idaho, Moscow

CABI (Center for Agriculture and Biosciences International) 2019 Invasive Species Compendium. https://www.cabi.org/isc. Accessed 1 Aug 2019

Carnahan MM, Goddard D, Lindelof D 2013 *World War Z*. Film based on the 2006 novel by Max Brooks

Carson R (1962) Silent spring. Houghton Mifflin, New York

Clark GH, Fletcher J (1909) Farm weeds of Canada, 2nd edn. Canada Dept. of Agriculture, Branch of the Seed Commissioner. Government Printing Bureau, Ottawa

Cocannouer JA 1950 Weeds, guardians of the soil. The Devin-Adair Company, Old Greenwich, CN. PDF file hosted by Peter Andrews at: http://www.naturalsequencefarming.com. Accessed 1 Aug 2019

© Springer Nature Switzerland AG 2020

G. Rogers, *Desert Weeds*, https://doi.org/10.1007/978-3-030-45854-6

Colautt RI, Richardson DM (2009) Subjectivity and flexibility in invasion terminology: too much of a good thing? Biol Invasions 11:1225–1229

Cole DN (1990) Trampling disturbance and recovery of cryptogamic soil crusts in Grand Canyon National Park. Great Basin Naturalist 50:321–325

Copple RF, Pase CP. 1967. A vegetative key to some common Arizona range grasses. U. S. Forest Service Rocky Mountain Forest and Range Research Station, Research Paper RM-27

Cox JR (1992) Lehmann lovegrass live component biomass and chemical composition. J Range Manag 45:523–527

CropTrust 2019. https://www.croptrust.org. Accessed 1 Aug 2019

Crosby AW (1952) The Columbian exchange: biological and cultural consequences of 1492. Greenwood Press, Westport

Di Castri F (1989) History of biological invasions with special emphasis on the old world. Pages 1-30 in Drake et al., eds. SCOPE 37, biological invasions, a global perspective. John Wiley & Sons, New York, 506 p.

Dillman AC (1946) The beginnings of crested wheatgrass in North America. Journal of the American Society of Agronomy 38:237–250

eFloras. http://www.efloras.org. Accessed 25 July 2019

EOL 2019 Encyclopedia of Life. https://eol.org. Accessed 25 July 2019

Frenkel RE (1970) Ruderal vegetation along some California roadsides. University of California Publications in Geography, University of California Press, Berkeley

Genesys 2019. https://www.genesys-pgr.org. Accessed 25 July 2019

Gillam C (2017) Whitewash, the story of a weed killer, cancer, and the corruption of science. Island Press, Washington

Gledhill D (1985) The names of plants. Cambridge University Press, Cambridge

Gori DF, Enquist CAF 2003 An assessment of the spatial extent and condition of grasslands in central and southern Arizona, southwestern New Mexico, and northern Mexico. The Nature Conservancy, Arizona Chapter. http://azconservation.org/downloads/multi/category/Arizona. Accessed 28 June 2019

Groh H (1940) Hoary cresses in Canada. Science of Agronomy 20:750–756

Heady HF (1977) Valley grassland. In: Barbour MG, Major J (eds) Terrestrial vegetation of California. John Wiley and Sons, New York, pp 491–514

Hendricks DM (1985) Arizona Soils. College of Agriculture, University of Arizona, Tucson

Hitchcock AS, Chase A (1950) Manual of the grasses of the United States, 2nd edn. USDA Misc. Publ, Washington. No. 200

Hobbs RJ, Humphries SE (1995) An integrated approach to the ecology and management of plant invasions. Conserv Biol 9:761–770

Holm LG, Plucknett DL, Pancho JV, Herberger JP (1977) The world's worst weeds. University Press, Honolulu

International Union for the Conservation of Nature (IUCN), Invasive Species Specialist Group 2019. http://www.issg.org/worst100_species.html. Accessed 1 Augs 2019

Jabran K, Chauhan B (eds) (2018) Non-chemical weed control, 1st edn. Academic Press, Cambridge

Jameson CM (2012) Silent spring revisited. Bloomsbury, London

Janzen DH (1967) Why mountain passes are higher in the tropics. Am Nat 101:233–249

Kane CW (2017) Medicinal plants of the American southwest. Lincoln Town Press, Oracle

Köhler FE 1883 Medizinal-Pflanzen in naturgetreuen Abbildungen mit kurz erläuterndem Texte Gera-Untermhaus, Germany. Artist unknown (Köhler?). Public domain

Leon-Sanchez L, Nicolas E, Goberna M, Prieto I, Maestre FT, Querejeta JI (2017) Poor plant performance under simulated climate change is linked to mycorrhizal responses in a semi-arid shrubland. J Ecol 105:960–976

Leopold A (1949) A Sand County almanac and sketches here and there. Oxford University Press, New York

Mabey R (2011) Weeds, in defense of nature's most unloved plants. Harper Collins Publishers, London
Mack RN (1981) Invasion of Bromus tectorum L. into western North America: An ecological chronicle. Agro-Ecosystems 7:145–165
Mack RN (1989) Temperate grasslands vulnerable to plant invasions: Characteristics and consequences. Pages 155–179 in Drake et al., eds. SCOPE 37, biological invasions, a global perspective. John Wiley & Sons, New York, 506 p.
Mack RN, Foster SK (2009) Eradicating plant invaders: combining ecologically-based tactics and broad-sense strategy. In: Inderjit (ed) Management of invasive weeds. Springer Science, New York, pp 35–60
Marchand PJ, Roach DA (1980) Reproductive strategies of pioneering alpine species: seed production, dispersal, and germination. Arct Alp Res 12:137–146
Marlette GM, Anderson JE (1986) Seed banks and propagule dispersal in crested-wheatgrass stands. J Appl Ecol 23:161–175
Martin DL, Bradley FM (2018) Rodale's ultimate encyclopedia of organic gardening: the indispensable green resource for every gardener. Rodale Press, New York
Masclef, Amédée. 1891-1893. Atlas des plantes de France. In *Librairie des Sciences Naturelles*, ed Paul Klincksieck, n.p. Paris, Sorbonne
Maxwell BD, Lehnhoff E, Rew LJ (2009) The rationale for monitoring invasive plant populations as a crucial step for management. Invasive Plant Science & Management 2:1–9
McReynolds K (2008) Invasive plants: sweet Resinbush. Rural living in Arizona 2:9
Mohlenbrock RH (1989) Midwest wetland flora: field office illustrated guide to plant species. Lincoln, Midwest National Technical Center, USDA NRCS Wetland Science Institute
Mohlenbrock RH (1995) Northeast wetland flora: field office guide to plant species. Northeast National Technical Center, USDA NRCS Wetland Science Institute, Chester
Moore MM (1989) Medicinal plants of the desert and canyon west. Museum of New Mexico Press, Santa Fe
Moore MM (2003) Medicinal plants of the mountain west. Museum of New Mexico Press, Santa Fe
Mueller-Dombois D, Ellenberg H (1974) Aims and methods of vegetation ecology. Wiley, New York
Myers JH, Bazely DR (2003) Ecology and control of introduced plants. Cambridge University Press, Cambridge
National Invasive Species Council 2006 Invasive species definition clarification and guidance white paper. https://search.usa.gov/search?query=Invasive+species+definition+clarification+and+guidance+white+paper.+&op=Search&affiliate=doi.gov. Accessed 28 June 2019
Native American Ethnobotany 2019 Native American Ethnobotany database. http://naeb.brit.org. Accessed 3 Aug 2019
Naylor RL (2000) The economics of alien species invasions. In: Mooney HA, Hobbs RJ (eds) Invasive species in a changing world. Island Press, Washington, pp 241–259
Parker KF (1972) An illustrated guide to Arizona weeds. University of Arizona Press, Tucson
Parvez S et al (2018) Glyphosate exposure in pregnancy and shortened gestation length: a prospective Indiana birth cohort study. Environ Health 17:23
Pauchard A et al (2009) Ain't no mountain high enough: plant invasions reaching new elevations. Front Ecol Environ 7:479–486
Phillips BG, Crisp D (2001) Dalmatian toadflax, an invasive exotic noxious weed, threatens Flagstaff pennyroyal community following prescribed fire. In: Maschinski J, Holter L (eds) Southwestern rare and endangered plants: proceedings of the 3rd conference; 2000 September 25–28; Flagstaff, AZ, tech. Coords. U.S. Department of Agriculture, Forest Service, Rocky Mountain Research Station, Fort Collins. RMRS-P-23: 200-205
Plants for a Future. 2019. http://www.pfaf.org. Accessed 1 Aug 2019
Qasem JR, Al Abdallat AM, Hasan SM (2019) Genetic diversity of *Solanum elaeagnifolium*, an invasive problematic weed in Jordan. Weed Res. https://doi.org/10.1111/wre.12360

Radosevich SR, Holt JS, Ghersa C (2007) Weed ecology: implications for management. Wiley, New York

Rapoport EH (1982) Areography: geographical strategies of species. Pergamon Press, New York

Reichenbacher FW (1984) Ecology and evaluation of southwestern riparian plant communities. Desert Plants 6:15–22

Reid CR, Goodrich S, Bowns JE (2008) Cheatgrass and red brome; the history and biology of two invaders. In: Kitchen SG, Pendleton RL, Monaco TA, Vernon J (eds) *Proceedings-Shrublands under fire: disturbance and recovery in a changing world; 2006 June 6–8; Cedar City, UT*, comps. U.S. Department of Agriculture, Forest Service, Rocky Mountain Research Station, Fort Collins, pp 27–32. RMRS-P-52

Ridley HN (1930) The dispersal of plants throughout the world. Ashford, Kent

Ripple et al (2017) World scientists' warning to humanity: a second notice. Bioscience 67: 1026–1028

Rogers, G (2019) Birds of Coldwater Farm, Dewey-Humboldt, Arizona. Coldwater Press, Prescott, AZ 177 pp.

Rogers RW, Lange RT (1971) Lichen populations on arid soil crusts around sheep watering places in South Australia. Oikos 22:93–100

Rogers, G., and J. Steele. 1980. Sonoran desert fire ecology: Adaptive strategies of perennial plant species. In *Proceedings of the fire history workshop, October 20–24, 1980*, tech coords. Marvin A. Stokes and John H. Dieterich. Fort Collins: U.S. Forest Service, General Technical Report RM 81:15–19

Salywon, Andrew, Dixie Damrel, and Wendy Hodgson. 2008. Agua Fria National Monument (in progress). http://swbiodiversity.org/seinet/checklists/checklist.php?cl=2624&pid=1. Accessed 29 June 2019

Schmidt A, Ibmayer J (1801-1809) Icones et descriptiones Graminum austriacorum. A. Schmidt, Vindobonae

SEINet. 2019. Environmental Data Portal. http://swbiodiversity.org. Accessed 25 July 2019

Shao MN, Qu B, Drew BT, Xiang CL, Miao Q, Luo SH (2019) Outbreak of a new alien invasive plant Salvia reflexa in Northeast China. Wiley, Hoboken. https://doi.org/10.1111/wre.12357. Accessed 29 June 2019

Sheldon JC, Burrows FM (1973) The dispersal effectiveness of the achene-pappus units of selected compositae in steady winds with convection. New Phytol 72:665–675

Simberloff D (2013) Invasive species, what everyone needs to know. Oxford University Press, Oxford

Slate ML, Callaway RM, Pearson DE (2018) Life in interstitial space: biocrusts inhibit exotic but not native plant establishment in semi-arid grasslands. J Ecol. https://doi.org/10.1111/1365-2745.13117. Accessed 29 June 2019

Southwest Biodiversity Program (2017). http://swbiodiversity.org/seinet. Accessed 29 June 2019

Southwest Monarch Studies (2019). https://www.swmonarchs.org. Accessed 29 June 2019

Southwest School of Botanical Medicine, Bisbee, Arizona. (2019). http://www.swsbm.com. Accessed 1 August 2019

Strum JG (1796) Deutschlands Flora in Abbildungen. https://doi.org/10.5962/bhl.title.507. Accessed 29 June 2019

SWEPC (2009). Southwest Exotic Plant Information Clearinghouse. http://sbsc.wr.usgs.gov/research/projects/swepic/swepic.asp. Accessed 29 June 2019

Tackenberg O (2003) Modeling long-distance dispersal of plant diaspores by wind. Ecol Monogr 73:173–189

Tackenberg O, Poschlod P, Bonn S (2003) Assessment of wind dispersal potential in plant species. Ecol Monogr 73:191–205

Tilford GL (1997) Edible and medicinal plants of the west. Mountain Press Publishing Company, Missoula

Tull D (2013) Edible and useful plants of the southwest: Texas, New Mexico, and Arizona. Univ. Texas Press, Austin

Turland NJ et al (2018) International code of nomenclature for algae, fungi, and plants (Shenzhen code), adopted by the nineteenth international botanical congress, Shenzhen, China. Koeltz Botanical Books, Eibelstadt

Turner RM, Webb RH, Esque TC, Rogers GF (2010) Repeat photography and low elevation fire responses in the southwestern United States. In: Webb RH, Boyer DE, Turner RM (eds) Repeat photography methods and applications in the natural sciences. Island Press, Washington, pp 223–244

U. S. Department of Agriculture Food Composition Databases (2019). https://ndb.nal.usda.gov/ndb. Accessed 25 July 2019

U. S. Department of Agriculture, Natural Resources Conservation Service (2019) Introduced, Invasive, and Noxious Plants. https://plants.usda.gov/java/invasiveOne. Accessed 29 June 2019

U. S. Department of Agriculture Plants Database (2019). http://plants.usda.gov/java/noxious-Driver. Accessed 29 June 2019

van der Pijl L (1972) Principles of dispersal in higher plants. Springer –Verlag, New York

Watson S (1871) Botany. Volume 5 in C. king. Report of the geological exploration of the fortieth parallel, vol 18. Professional Papers of the Engineer Department, U. S. Army, Washington. i-Liii, 1-525

Watson S (1874) Atriplex wrightii. Proceedings of the American Academy of Arts 9:113

Whitson TD, Burrill LC, Dewey SA, Cudney DW, Nelson BE, Lee RD, Parker R (2006) Weeds of the west. Western Society of Weed Science, Las Cruces

Wilson JS (2015) The bees in your backyard: a guide to North America's bees. Princeton University Press, Princeton

Wilson EO (2016) Half-earth: our planet's fight for life. W.W. Norton, London

Wuebbles DJ, Fahey DW, Hibbard KA, Dokken DJ, Stewart BC, Maycock TK (eds) (2017) Climate science special report: a sustained assessment activity of the U.S. Global Change Research Program, U. S. Government Printing Office, Washington

WWF (World Wildlife Fund) (2016) Living planet report 2016. Risk and resilience in a new era. WWF International, Gland

Yasin M, Rosenqvist E, Jensen SM, Andreasen C 2019. The importance of reduced light intensity on the growth and development of six weed species. Weed Research, Wily Online Library https://doi.org/10.1111/wre.12352. Accessed 29 June 2019

Young JA, Evans RA, Major J (1971) Alien plants in the Great Basin. J Range Manag 25:194–201

Zimdahl RL (2010) A history of weed science in the United States. Elsevier, New York

Zouhar K, Smith JK, Sutherland S, Brooks ML (eds) (2008) Wildland fire in ecosystems: fire and nonnative invasive plants. Rocky Mountain Research Station, Ogden. RMRS-GTR-42-Vol. 6

Index of Illustrators

© Springer Nature Switzerland AG 2020
G. Rogers, *Desert Weeds*, https://doi.org/10.1007/978-3-030-45854-6

Subject Index

© Springer Nature Switzerland AG 2020
G. Rogers, *Desert Weeds*, https://doi.org/10.1007/978-3-030-45854-6

The manufacturer's authorised representative in the EU is Springer
Nature Customer Service Centre GmbH, Europaplatz 3, 69115 Heidelberg,
Germany. If you have any concerns regarding our products, please
contact ProductSafety@springernature.com

Printed and bound by CPI Group (UK) Ltd, Croydon, CR0 4YY
28/09/2026
02224086-0013